James Edmund Harting

Rambles in Search of Shells, Land, and Freshwater

James Edmund Harting

Rambles in Search of Shells, Land, and Freshwater

ISBN/EAN: 9783744662109

Printed in Europe, USA, Canada, Australia, Japan

Cover: Foto ©berggeist007 / pixelio.de

More available books at **www.hansebooks.com**

RAMBLES

IN

SEARCH OF SHELLS,

LAND AND FRESHWATER.

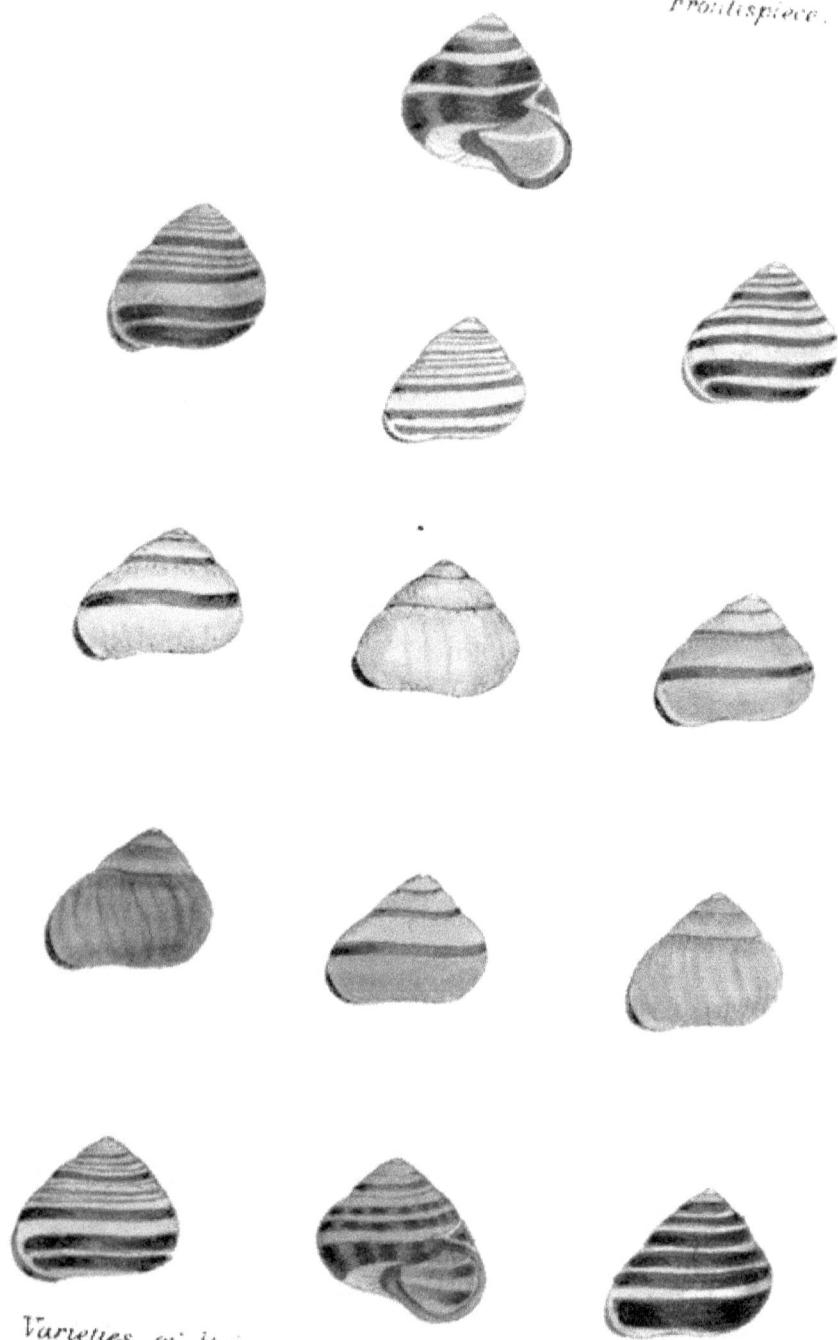

Frontispiece.

Varieties of Helix nemoralis, hortensis and hybrida.

RAMBLES

IN

SEARCH OF SHELLS,

LAND AND FRESHWATER.

BY

JAMES EDMUND HARTING, F.L.S., F.Z.S.,

AUTHOR OF A " HANDBOOK OF BRITISH BIRDS,"
" THE ORNITHOLOGY OF SHAKSPEARE,"
ETC. ETC. ETC.

With Coloured Illustrations.

LONDON:

JOHN VAN VOORST, PATERNOSTER ROW.

MDCCCLXXV.

LONDON :
PRINTED BY WOODFALL AND KINDER,
MILFORD LANE, STRAND W.C.

TO THE READER.

The following chapters, entitled " Rambles in Search of Shells," were originally published in the Natural History columns of " The Field," during the autumn of 1873 and the spring of 1874, and, judging by the inquiries which have since been made for a reprint, they appear to have found some little favour with naturalists.

Through the courtesy of the Proprietors of " The Field," to whom the Author here desires to express his obligations, these chapters are now republished, with some important additions and emendations, and with a series of coloured plates, which it is believed will materially assist the reader in the determination of the species mentioned or referred to in the course of the volume.

In penning these chapters, the Author has essayed to impart a little practical information on the subject of British land and freshwater shells without being too technical or systematic. Minute descriptions of form and colour have been avoided, as tending rather to perplex than assist the tyro in Conchology; and, in lieu thereof, an attempt has been made to give only such a description of each species as will secure its identification, some reliance, at the same time, being placed upon the distinguishing characters, as pointed out from time to time, of such nearly allied forms as are most likely to puzzle the collector.

The notion of grouping the species according to the soils they frequent, and the situations in which they are found, will, it is conceived, render the chapters more attractive than if they had been described seriatim in the order generally adopted by systematists.

In presenting this little work to the reader, the author feels that his acknowledgments are due to the late Dr. J. E. Gray and Mr. Gwyn Jeffreys, for their friendly assistance in looking over the proof sheets, and to Mr. J. Weaver, of Uppark, Sussex, for several interesting communications which have been embodied in the Introduction.

The plates have been carefully drawn and coloured by Mr. Arthur Rich, in every case from recent specimens—a merit not invariably possessed by modern illustrations to works on Natural History.

CONTENTS.

———

INTRODUCTION.

CHAPTER I.

CHAPTER II.

CHAPTER III.

CHAPTER IV.

CHAPTER V.

CHAPTER VI.

CONCLUSION.

RAMBLES

IN

SEARCH OF SHELLS.

—◆—

INTRODUCTION.

SNAILS and slugs in the abstract are not very
attractive objects to the million, at least this may
be said of the few species found in this country;
but if we were to limit our observations to those
natural productions only which have a pleasing
exterior, how many of the most interesting pages
of Nature's great book of wonders should we
pass over unread! How little, for instance, should
we know of the commonest of our molluscous
animals, beyond the generic and specific character
of the habitations of the shell-bearing species—

B

except that they remorselessly devour our cabbage
and other cultivated plants, or disfigure them with
their slimy trails as they crawl over them—if the
comparative anatomist, undaunted by their repulsive
appearance, had not by means of skilful dissection
learnt something of their wonderful structure and
given us the result of his investigations. In the
early days of conchology, it was held sufficient to
study the shells only of these animals, and the
possessor of an extensive collection of such shells
might be intimately acquainted with the name,
geographical distribution, and proper place in a
systematic arrangement of every specimen in his
cabinet without necessarily knowing anything of the
animal that formed it. Now, however, the conchologist has given place to the malacologist,
who, not content with examining, describing, and
naming the shell, independently of its inhabitant,
curiously questions the latter as to its habits and
internal structure, and in the case of those which
possess a single shell (Univalves), he literally learns
the relationship of each species from the animal's
own mouth.

Snails and slugs both have the power of drawing

in their horns on being touched, and this is effected
by a singular and beautiful apparatus; the tentacle
is lengthened by gradually unfolding itself, and not
by being pushed out from the base.   Each tentacle
is a hollow cylinder, to the apex of which is attached a muscle, arising from the retractor muscle
of the foot, and by its contraction the tentacle is
simply inverted and retracted, like the finger of a
tight glove; its protrusion, on the other hand, is
effected by the alternate contraction of the circular
bands of muscles which compose the walls of the
tentacle.   As a rule, slugs and snails are more
liberally provided with teeth than any other animals
in the parish, one of our slugs, for instance, pos-
sessing no less than 28,000; they are not, however, all in use at the same time.   The dental
apparatus of our univalves may be described as a
tube lined with teeth set upon flattened plates,
collectively called the lingual ribbon.   One extremity
of this ribbon is open and spread out like a tongue,
teeth upwards, on the floor of the mouth, so as to
occupy, in fact, the same relative position as the
tongue in the mammalia; the roof of the mouth is
supplied with a horny plate, against which the open

end of the ribbon can work backwards and forwards, so as to rasp and triturate the food between them. The tubular portion of this lingual ribbon is contained in a cavity behind the mouth, and as the teeth in use become worn or broken, it is conjectured that they are absorbed, and a fresh set from the reserve in the tube is pushed forward to take their place.

The body of every mollusk with which we are concerned, except those of the slug family, is contained in a membranous sac called the mantle, which not only serves as a model on which the shell is moulded, but is liberally provided at the edges of its open end with the glands that secrete the shelly matter. To this set of glands alone are due the coloured bands and other markings in the shells, as may be seen in the case of a fractured specimen that had been repaired by the owner; in this, the new matter thrown out of the mantle under the fracture is always colourless.

Some curious observations on the growth of shells of land snails were communicated by Mr. E. J. Lowe, to the Royal Society in 1854. He found that the shells of *Helices* increase but little for a considerable period, never arriving at maturity before the animal has once become dormant. The shells, it would appear, do not grow whilst the animal itself remains dormant, but the growth is very rapid when it does take place. Most species bury themselves in the ground to increase the dimensions of the shells; and in illustration of this Mr. Lowe states that a pair of *Helix aspersa* had deposited their eggs, which began to hatch on the 20th of June. The young ones grew but little during the summer: they buried themselves in the soil on the 10th of October, coming again to the surface on the 5th of April, not having grown during the winter. In May they buried themselves with their heads downwards (in winter they and other species buried themselves with the heads upwards) appearing again in a week, double the size. This process was carried on at intervals of about a fortnight until the 18th of July, when they were almost fully grown.

The process of growth within the ground takes place with *Helix nemoralis*, *H. virgata*, and *H. hispida*. But *H. rotundata* burrows into decayed wood to increase the size of the shell: whilst *Zonites radiatulus* appears to remain on decaying blades of grass; and *Pupa umbilicata, Clausilia rugosa,* and *Bulimus obscurus* bury their heads only to increase their shells. With respect to *Zonites cellarius, Z. nitidus,* and *Z. nitidulus,* it was not satisfactorily ascertained whether their heads were buried during the process of growth or not.

We are all tolerably familiar with the fact that one essential character of the vertebrated animals is their possession of a brain and spinal cord, from which proceed those

> "Tracts along which the mysterious will
> Is conveyed, by a process on which Fancy lingers
> With awe, from the brain to the tips of the fingers,"

or their analogues, and other portions of the animal frame. In the species under consideration this plan is considerably modified, and instead of one cerebral mass supplying the whole system, we find half a dozen ganglionic centres from which the nervous threads radiate to the organs of sight, smell, hearing, and touch (which appear to be well developed), and the important systems of digestion, circulation, respiration, locomotion, and reproduction.

Respiration in the animals of this class is carried on by means of a rudimentary lung in each member of the terrestrial division ; the aquatic species, with some exceptions, being provided with gills. In the air-breathing molluscs, the air is admitted into the pulmonary cavity, not by the mouth but by an aperture, which may easily be seen in slugs at the edge of the convexity on the back formed by the mantle, and in snails just within the mouth of the shell ; in both cases on the right side.

In regard to locomotion, most univalves crawl upon a large fleshy protuberance which is the homologue of a foot and supports the body, and many, even amongst the bivalves, by means of this large foot, are enabled to traverse considerable distances. Not a few of the aquatic univalves are able to swim, or rather creep, upside down upon the under surface of the water.

The mode of reproduction amongst mollusca varies. Many univalves have distinct sexes, while most of the land snails are hermaphrodite. Some, like the *Valvatidæ*, change their sex after a time, being at first male, and then female. The majority are oviparous, but some are ovoviviparous, and their wonderful fertility may be estimated from a statement of Pfeiffer to the effect that the gills of an

average sized freshwater mussel contain about
400,000 eggs.

If we add to this faint outline of the internal
structure of snails and slugs, that these animals are
essential in checking the redundancy of vegetation,
removing decomposing matter, both animal and vegetable, and supplying dainty food to many other
members of the animal kingdom, we shall have said
enough, perhaps, to show that they possess a high
degree of interest.

It has often been a matter of surprise to us that
the study of the land and freshwater shells has not
more votaries, especially amongst the fair sex. The
subject may be easily coupled with botany, being,
as it were, nearly associated with it; for whether
we ramble on the downs, in the woodland, or in
the marsh, in search of any particular plant, we
seldom fail to find in close proximity to it some
species or other of mollusca which claims its shelter
or support.

CHAPTER I.

THE number of British land and freshwater
mollusca is somewhere about one hundred and
twenty, excluding such as have been evidently
introduced, or described as British on doubtful
authority, but including the slugs, which, though
generally regarded as shell-less, have the shell
either rudimentary, and of an indefinite form, or
shield shape, placed beneath the mantle.

Of these one hundred and twenty, about seventy-
five are terrestrial, and forty-five aquatic; but it
is, of course, to be understood that, in speaking
of them as British, we do not mean to imply that
they are not found out of the British Islands.
So far from this being the case, the majority of
them are commonly distributed over the north of
Europe, while many are found not only in Southern
Europe, North Africa, and Asia Minor, but even

in Siberia and the northern parts of Asia and
America. Of these we may have occasion to
speak later. Meantime it will be desirable to
get a general idea of some of the commoner forms
of shells, and the subject is at once simplified by
dividing the whole number of species into two
classes—the Bivalves (*Conchifera*), of which the
common mussel furnishes an illustration, and the
Univalves (*Gasteropoda*), of which the garden snail
is a familiar example.

Amongst the aquatic shells there are both bivalves
and univalves, but with the terrestrial species only
univalves occur. The reason for this is to be found
in the difference of structure which exists in the
animal itself, particularly in the organs of respiration, and which enables one species to breathe and
live where another would assuredly die.

The bivalves, which are all *Lamellibranchiata*,—
that is, having leaf-like gills,—may be separated into
three very distinct and well-marked families—namely,
Sphæriidæ (from the spherical shape of the species) ;
Unionidæ, which contains the mussels (from *unio*,
a pearl, one of them being a pearl-bearing species),
and *Dreissenidæ* (so named after M. Dreissens, a

Continental naturalist), of which family, however, we have but a single genus and species, the well-known "zebra mussel."

The aquatic univalves are either *Pectinibranchiata* —that is, having comb-like gills; or *Pulmonobranchiata*—those with lung-like gills. In the former order there are three families, which take their names from some of the genera which they contain —*i.e.*, *Neritidæ*, *Paludinidæ*, and *Valvatidæ*; in the latter order, one large family, *Limnæidæ*, contains the genera *Limnæa*, *Physa*, *Planorbis*, and *Ancylus*.

The terrestrial univalves comprise three families, *Limacidæ*, *Testacellidæ*, and *Helicidæ*, which in plain English are known as the slugs, the shell-slugs, and the snails, with two much smaller families, to one of which no English name has been given, and which have been formed each for the reception of a single genus (*Carychium* and *Cyclostoma*) of peculiar character.

All this may appear very dry to the unscientific reader; but it is scarcely to be dispensed with, since without a system of some sort it would be not only impossible to arrange and store one's facts, but equally impossible to institute comparisons and draw conclusions.

Without the aid of figures it is very difficult to give an accurate idea of shape and size ; yet some knowledge of the kind is requisite before we can make any progress.

Let us see how far a verbal description will answer, if confined to species which may be regarded as mere modifications of a common type or standard. The idea, we believe, has not been previously suggested.*

Take the shell of the common garden snail (*Helix aspersa*) as a standard. To this genus (*Helix*) belong about five-and-twenty species out of the seventy-five land shells which are generally regarded as British. We shall consider the distribution and rarity or otherwise of these twenty-five later.

Now, imagine this type or standard to be in a soft and plastic condition, and capable of being modelled. If, holding it by the lip with the left hand, and taking the apex of the spire between the finger and thumb of the right hand, we slightly elongate it, what is

* The reader may be reminded that when this chapter was first published no illustrations accompanied it.

the result? We increase the length, decrease the width, and cause the whorls to become more diagonal than horizontal. Deepen the sutures, and we obtain in effect a rough model of another genus, *Paludina*, of which we have two species—*P. contecta*, with a glossy shell of a yellowish or greenish-brown colour, with bands and striæ of a darker brown; and *P. vivipara*, which is somewhat smaller, less glossy, the whorls less swollen, the sutures less deep, and the mouth less circular. Both species inhabit ponds and rivers.

If we now elongate and attenuate the spire, lessen the depth of the sutures, and imagine the shell to have become so much thinner as to be almost transparent, we get a notion of *Limnæa*, a widely-distributed, marsh-loving genus, of which there are some eight species in Great Britain. They are all found in still and shallow waters, the best known amongst them being *L. stagnalis, auricularia, peregra* and *palustris.* The reproductive system of the *Limnæidæ* and other families of the order *Pulmonobranchiata* is very singular. The individuals of species which possess an operculum are of different sexes, while in those which have no operculum the sexes are united in the same individual. In the latter class, however, an individual requires to be fertilized by another of its own species. But, as observed by Leach, the sexual parts are so far situated from each other, that one individual is able to perform the functions of each sex with two others at the same time, and it is consequently not uncommon to find several of these molluscs engaged at once in a mutual interchange of attentions.

Having thus changed the shape of our model, let us now reduce the size of it, increase its elongation, and narrow the width until the shell becomes oblong, with the whorls drawn out and spire long. We then get some idea of another genus, *Bulimus*, of which there are three British species, all herbivorous —*acutus*, the commonest, inhabiting downs and sand-hills ; *montanus* (or *lackhamensis*) and *obscurus*, both of which are found on the trunks of trees and amongst dead leaves in woods and hedgerows.

By cutting off a couple of whorls or so from the apex, and rounding it, at the same time compressing the model so as to make the whorls more compact, we get some notion of *Pupa*. The four species in this genus—*secale, ringens, umbilicata*, and *margin-*

ata—are all of small size and gregarious habits, living in moss or amongst stones and roots. Besides their variation in size and colour, they are to be distinguished by the curious processes called "teeth," which fence in and contract the mouth of the shell.

The eight or nine species which compose the closely allied genus *Vertigo* are but miniature forms of *Pupa*, and their habits are similar. The chief difference is to be found rather in the animal than in the shell, the inhabitant of which has two tentacles instead of four, as in *Pupa*.

If we now restore the pointed apex which we cut off from *Bulimus* to model *Pupa*, and, compressing the shell laterally, elongate it until the length is five or six times greater than the width, we have something like *Clausilia* before us. It is a spindle-shaped shell, with a longer spire than *Bulimus*, and is especially characterized by having a curious internal process called the "clausilium" (hence the name), which acts as a valve or lid in closing the shell against all intruders. There are four species to be met with in the British Islands—*C. biplicata, laminata, rugosa,* and *Rolphii*—all of which frequent the trunks and roots of trees, and may be found also on moss-covered stones and amongst dead leaves and *débris* in woods.

Having so far stretched our model to its utmost while reducing it in size and width, let us now restore it to the shape in which we found it, that is, to the shape of a common *Helix*, and attempt some modifications in another direction. Instead of elongating let us compress the shell vertically, flattening the spire and all the whorls until the shape is that of an ammonite, or, to use a more familiar simile, a catharine-wheel. We have then a rough notion of what the genus *Planorbis* (*i.e.*, flat coil) is like. In this genus there are said to be eleven British species, although it requires a nice discrimination to identify what are, except in regard to size, but eleven slight modifications of the common type of which we have endeavoured to give some outline.

The largest of them is *Planorbis corneus*, being in diameter, when adult, about the size of a shilling, and in thickness about a quarter of an inch. The others are all very much smaller and flatter, varying in depth from the thickness of a penny to that of a knife-blade. This genus has some remarkable peculiarities, which have been well pointed out by

Mr. Gwyn Jeffreys in the first volume of his
" British Conchology." One of them is the habit
which the animal has of emitting its purple-
coloured blood on being irritated, apparently as a
means of defence. This is especially noticeable
from its large size in *P. corneus;* and although
experiments have been made with the view of fix-
ing and utilizing the purple dye thus yielded, they
have hitherto proved unsuccessful. Another peculiarity is, that several of the vital organs are placed
on the left side of the body, instead of on the
right (as in most other univalves), while the shell
is coiled the other way, that is, from left to right.

Some of the smaller species, found in marshy
ground which becomes dried up in summer, close
the mouth of their shell with an epiphragm or filmy
covering, and live in retirement until a return of
moisture tempts them forth again. They are all
herbivorous in their nature.

It would not be possible within the limits of the
present chapter to attempt a description of all the
genera of land and freshwater mollusca. In the
above remarks we have directed attention only to
the more prominent forms. We have not described

Zonites, which is nearly related to *Helix*, nor *Physa*, which is close to *Limnæa*, and have passed over many other genera, which, though allied to others above mentioned, might, if named, confuse perhaps rather than instruct.

We shall ask the reader, later, to accompany us in spirit in a ramble in search of shells over the London clay; and in a succeeding chapter over the chalky downs and through the moist beech woods of Sussex —ground most fertile to the conchologist. We shall then search for the prettily spotted *Neritina fluviatilis*, the delicately coloured *Cyclostoma elegans*, the edible snail (*H. pomatia*) and the freshwater limpets (*Ancylus*), which, with many other species, reward with unspeakable pleasure the collector who finds them for the first time.

The mention of limpets reminds us of that beautiful little bird the Dipper or Water-ouzel, which feeds largely on these and other freshwater molluscs; and we may here remark that, as a variety of small shells may be found in the stomachs of many species of birds, the ornithologist who cares to take up conchology may thereby add much to the knowledge which he already possesses on his favourite subject.

CHAPTER II.

IN the last chapter we promised to say something of the land and fluviatile mollusca inhabiting the London clay. Those who possess gardens in the metropolis and its environs may not be prepared, perhaps, to learn that they may commence a collection of shells before leaving their own premises ; yet such is the case. There are certain species so generally dispersed, or of such an accommodating nature, as to find an existence in that generally barren territory where the only birds are sparrows, and the indigenous animals cats. Others, no doubt, get introduced with plants and shrubs, and ultimately become, so to say, acclimatized.

The commonest species naturally presents itself first to our notice ; but in referring to this, the ordinary garden snail, *Helix aspersa* (Pl. I., fig. 3), which must be sufficiently familiar to all, it will only be necessary to do so in order to point out a few facts

c 2

in connection with it which may not be so generally
known.   Cowper says :

> " Who seeks him must be worse than blind,
>   He and his house are so combined,
>   If finding it he fails to find
>   Its master."

And it is about the "master" rather than the "house"
that we have now to speak.   Of the destructiveness
of this mollusc to fruit and vegetable we need say
nothing, but perhaps it is not generally known to be
carnivorous.   Not only will it greedily feed upon
such fresh meat as it can reach, but it has been
known to kill and partially devour a large black slug
which had been previously confined in the same
vessel with it.

Per contra, its use in medicine and as an edible
delicacy, for those who like it, should be taken into
account.   For both these advantages we are indebted
to our French neighbours, who have not only shown
us the efficacy of " Hélicine " in cases of whooping
cough and other pulmonary complaints, but, through
that culinary art for which they are renowned, have
illustrated the proverb which says that the best
appeal to an Englishman's heart is through his
stomach.

1

2

3

4.

5

6.

7.

8.

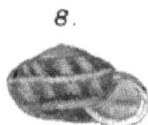

1. Helix hortensis 2. H. nemoralis 3 H. aspersa.
4. H pulchella 5. H. rotundata. 6. H. arbustorum.
7. H. hispida. 8. H. rufescens.

A. Rich lith

W. West & Cº imp

As a probable independent recognition of the snail's use in medicine, we may mention that the village dames in Sussex, even at the present day, recommend " snail syrup " as a specific in all cases of cough and cold.   Their mode of preparing it is as simple as it is said to be efficacious.   A score or so of snails are strung together by means of a needle and 'stout thread, which is passed through the shell and body of each.   They are then suspended festoon-like over a dish or pan of coarse brown sugar, on which the mucilaginous fluid is allowed to drop. The resulting compound is a syrup of snails, of which two teaspoonfuls twice a day is said to be the proper dose.

This glutinous exudation was formerly used for bleaching wax, and, in part, for making cement, until other less troublesome methods came to be employed.

Few, perhaps, are aware that the snail is capable of producing musical sounds; but the fact has been thus described to us by an observant friend.   One stormy evening in autumn, while engaged with his books in a room, the windows of which were directly exposed to the wind and rain, he was startled by a

sound so sweet—so unlike anything he had ever
heard before, and so peculiar—that he was not
only puzzled to imagine a probable cause for it, but
actually unable to form a guess as to the exact
spot whence it proceeded. It was often repeated at
irregular intervals during the evening, occasionally
three or four times in the course of as many
minutes; and, although it was not loud, and did not
continue many seconds at a time, it was distinctly
audible, notwithstanding the noise occasioned by
the beating of the rain against the windows and
the discord of the wind in the chimney stacks.
Several months elapsed ere he had an opportunity of
hearing it again. At this second performance the
rain was falling as before, but the shutters of the
room were not closed as they had been in the first
instance, and he was enabled distinctly to trace the
mysterious sound to one of the windows. A light
held near this window revealed the fact that the
musician was no other than our friend the *Helix
aspersa*. It appeared that the friction of its foot or
shell—he could not satisfactorily determine which—
against the wet glass had caused a vibration in the
pane on which it was travelling, and in this way

produced a note not unlike that which we all know may be brought out of a drinking glass by the friction of a wet finger on the rim.

Snails are very sensible of cold, and commence to hybernate early, clustering together in the crevices of old walls and trees, or behind water-butts and flower-boxes, and attached to each other by the epiphragms which close the mouths of their shells. Some notion of their prolific nature may be gathered from the statement of a French naturalist, who affirms that he has counted upwards of a hundred eggs laid by a single individual. In this respect, however, the common garden snail, *Helix aspersa*, excels most of its congeners.

Mr. H. Adams informed the Editor of the "Zoologist" that he once found a rarity in the shape of a reversed specimen of this species in his garden at Notting Hill.

Amongst the other species of snail to be met with in gardens may be mentioned the shrub snail, *Helix arbustorum;* the little banded garden snail, *H. hortensis;* and the wood snail, *H. nemoralis.* The shrub snail (Pl. I., fig. 6) is much smaller than the common garden snail, with a smoother and more glossy shell, prettily mottled with wood brown upon a lighter ground colour. The shell is also more solid, and has the lip thick, white, and reflected. It cannot, however, be considered common in gardens, at least in London, for it requires more moisture and shade than is generally found there, and is more frequently met with in copses, amongst nettles, and upon alders by the river side. It has been found, nevertheless, at Hammersmith, Fulham, Charlton, and Battersea. We have referred to *H. hortensis* and *H. nemoralis* as distinct species, but whether they are so has long been a matter of controversy amongst conchologists. Linnæus united them; Müller separated them. In modern times Messrs. Forbes and Hanley agree with the former, and Dr. Gray with the latter. Mr. Norman contends stoutly that they are not the same species, and his principal reason is that *nemoralis* invariably, but *hortensis* never, has a calcareous and frequently coloured deposit on the columella. We have been hitherto inclined to take Mr. Norman's view, in further confirmation of which we have noticed that *nemoralis* always has a black lip (Pl. I., fig. 2), *hortensis* a white one (Pl. I., fig. 1). But Mr. Gwyn Jeffreys—to whom all conchologists are indebted for his able and most instructive work on British conchology—holds that Linnæus was right, and observes that the variety *hybrida* (Frontis., fig. 12) seems to connect the two forms so far as concerns their conchological distinction ; and the only character of importance upon which a difference between them can be founded, consists in a slight variation of shape in their love-darts.* He accordingly regards *H. nemoralis* as the type, and *hortensis* and *hybrida* as local or casual varieties of one and the same species. He has never found any two of these forms living together, and other observers have made the same remark. There are endless varieties of each. (*See Frontispiece.*)

Wherever any rank growth of nettles and other weeds has been allowed to stand, whether in gardens or elsewhere, we may look for the little *Helix*

---

* During the pairing season, snails of this genus are furnished with little crystalline darts, which, after many little coquettings, they shoot out towards each other. They are contained in a special pouch or receptacle ready for use, and their shape varies according to the species. Some individuals have only one, others two, while in a few species they are wanting altogether. After such conflicts, these curious love-darts may be found sticking in the bodies of the wounded.

rufescens (Pl. I., fig. 8). It is of a reddish-grey or brown colour, closely striated transversely, and rather solid and opaque, at the same time much more compressed in form than any of those before mentioned. This and the two last-named species (or species and variety, as we suppose they must now be called) furnish plenty of food to the blackbirds and thrushes. The latter, in particular, are very fond of them, and may be seen searching for them with great perseverance in many a weed-choked ditch. Little heaps of empty shells, with the spires broken, may often be found in our gardens, testifying to the feast which has rewarded the industrious songster.

In similar situations, as well as under logs and stones, may be found the little bristly snail shell, *H. hispida* (Pl. I., fig. 7), which is not unlike *rufescens* in size and colour, but with a thick epidermis, closely covered with short recurved hairs, which are persistent and not easily rubbed off. It may be observed, by the way, that the young of *rufescens* have the shell also hispid. The hairs may be easily seen with a lens, and after they have fallen off, the impressions which are caused by their insertion into the epidermis remain on the surface of full-grown specimens, and may be easily discerned under the microscope.* These small snails, which are often very destructive in gardens, lay from forty to fifty eggs in August and September, from which the young are excluded in about three weeks. *H. hispida* has been found under stones in Hyde Park.

Mr. Gwyn Jeffreys has described another species, *H. concinna*, closely allied to the last named, but differing in having a more glossy shell, which is never globose like *hispida*, and has the umbilicus more open, with the hairs more scattered and more easily shed. The animal, too, is of a darker colour, with a narrower and less fleshy foot.

A smaller species still is *H. rotundata* (Pl. I., fig. 5), which may be found under stones and bark, and amongst moss and dead leaves, and is everywhere tolerably common. The shell, as its name implies, is nearly circular, more compressed below than above, rather thin, but nearly opaque, and moderately glossy. It is of a yellowish-brown, or horn colour, transversely marked with reddish-brown streaks.

* See a note by Captain Bruce Hutton in the "Zoologist," 1862, p. 7977.

Under stones, about rockwork in gardens, and in
moist situations generally, may be found the little
Helix pulchella (Pl. I., fig. 4), a solid but transparent and glossy shell of a light grey or white colour.
Amongst the localities on record for this widely distributed species are gardens at Chelsea, Hammersmith, Blackheath, and Eltham. This and the four
last named, from their small size, require to be more
carefully looked for than the brighter coloured and
more obtrusive banded snails.

The genus *Helix* comprises all the true snails,
which have shells more or less globular, and usually
a semilunar mouth. They also have the teeth
notched or serrated. In the closely allied genus
Zonites the animals have the same glassy-looking
shells as in *Vitrina*, and, being of much the same
habits, have similarly hooked teeth. They frequent
dark, damp spots, generally under stones, old bricks,
and logs of wood partly buried, as well as amongst
dead leaves and moss. *Zonites cellarius* (Pl. II.,
fig. 1) is found in cellars, drains, and sculleries, and
under tiles or loose bricks about houses. The shell
is thin and brittle, but very glossy and semitrans-
parent, and of a yellowish or brownish horn colour.

1

2.

3

4.

5.

7

6

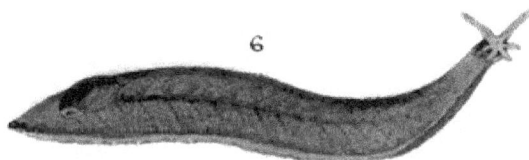

1. *Zonites cellarius.* 2. *Pupa secale.* 3. *P. marginata*
4. *Zonites nitidulus.* 5. *Vertigo antivertigo.* 6. *Testacella*
haliotidea. 7. *Shell of ditto.*

A. Bick lith   W. West & Co imp.

Another kind of *Zonites*, namely, *alliarius*, with a darker and more solid shell, has a very strong smell of garlic (hence the specific name), especially when irritated. This peculiar smell, however, varies in intensity, and is sometimes hardly perceptible, even when the animal has been much provoked. A third species, *nitidulus* (with a variety *nitens*), is not uncommon on the banks of the Thames near London, preferring more watery places than the last named (Pl. II., fig. 4). It has been met with also at Camberwell and Hammersmith. In similar situations occurs *Z. nitidus*, a variety of which, named *albida*, has been found among the rejectamenta of the Thames at Richmond. Dr. Gray has obtained *Z. crystallinus* at Battersea. It is a thin, glossy, and transparent shell, of a greenish-white colour and glassy appearance, and the animal inhabiting it is of a clear greyish-white colour, and nearly transparent. Another species of *Zonites* (*Z. glaber*) has lately been found in Hertfordshire, and some other parts of England. It somewhat resembles *Z. alliarius*, but is much longer, and has more convex or swollen whorls.

The minute shells of the genus *Pupa* are not very readily seen; and were it not for the fact that the

species in this genus are gregarious, and attract
more attention when clustered together than they
would do singly, we might look a long time before
finding them. The only two which, so far as we are
aware, have been found in or near London, are *Pupa
umbilicata* and *marginata, vel muscorum,* (Pl. II.,
fig. 3), both of which are partial to old walls and
roots, and may be looked for with a good chance of
success about the artificial rock or grotto work
with which many fern-growers are wont to ornament
their gardens. Mr. Rich informs us, however, that
Pupa secale, vel juniperi, (Pl. II., fig. 2) has been
met with on an old wall at Sudbury, near Harrow.

We do not remember to have seen any species of
Vertigo from London gardens, although some of them
are far from rare, and, being of similar habits, are to
be found in similar situations to *Pupa,* of which
genus they may be said to be miniature forms. The
animal of *Vertigo,* however, has but one pair of tentacles, while *Pupa* has two pairs. The spire of the
shell is shorter, and the outer lip more contracted.

To find some of the commoner kinds of *Clausilia*
and *Bulimus,* we must go to the woods and fields ;
for it would be vain to search the smoke-dried and

soot-begrimed trees which in purer air and soil would be the haunt of these and other woodland species.

Before leaving the garden, however, we may look for the curious shell-slug, *Testacella haliotidea* (that is, resembling a *Haliotis* or ear-shell), which is not very rare in London gardens, although, from its habit of burrowing, it is often overlooked. The metropolitan form, however, is said to differ sufficiently and permanently from *haliotidea* to warrant its being regarded as a permanent variety, and it was accordingly described by the late Mr. George Sowerby as *Testacella scutulum*. To this variety probably belongs the animal described by Mr. Tapping (" Zoologist," 1856, p. 5105) as *Testacella Medii Templi*, from its having been found under the shelter of a south-west wall, in the Middle Temple Gardens. Specimens have been procured from time to time in the Botanic Gardens, Regent's Park; in the Circus Road and Adelaide Road, St. John's Wood; at Hampstead, Hendon, Kensington, Hammersmith, and Lambeth (Pl. II., fig. 6, 7).

The animal partakes of the nature both of a slug and a snail, having a long naked body, and a comparatively small and flat shell, which serves to protect the heart, liver, and other vital organs. In the colour of the body and size of the shell it varies very much ; hence varieties have come to be described as new species. It is said to be the only land mollusc which has truly predaceous habits, feeding on earthworms, which it pursues under ground, and devouring snails, slugs, and even others of its own species. Dr. Ball writes : " I first became aware of this *Testacella* preying on worms by putting some of them in spirits, when they disgorged more of these animals than I thought they could possibly have contained ; each worm was cut, but not divided, at regular intervals. I afterwards caught them in the act of swallowing worms four and five times their own length."

When following the worm through its winding tunnels, the *Testacella* finds in its small flat shell a useful defence against similar attacks upon itself from the rear, for as it moves along, the shell serves to block up the passage, and at the same time acts as a shield by which the whole body is guarded. In dry weather this slug retires into a sort of nest or cocoon, formed of slime, which gradually dries and hardens, and in this it remains in a state of semi-torpor, until more genial weather tempts it forth again.

An excellent account of this curious animal, too long to be quoted here, will be found in the first volume of Mr. Gwyn Jeffrey's " British Conchology," p. 141 ; and to this volume we may also refer such of our readers as desire more minute details regarding the species mentioned in these brief and necessarily very general remarks. Considering the many different shells which are deserving of notice, it is not possible in the space at our disposal to do more than indicate, as we have attempted to do, their general appearance and peculiarities, with some hints as to the localities they frequent, and the situations in which they are most likely to be found.

In our next chapter we shall proceed from town to country, where we may expect to find not only better specimens of such shells as are procurable in gardens, but many others—both terrestrial and aquatic—which will be new to the collector.

CHAPTER III.

HAVING made acquaintance with the various species
of mollusca to be met with in and about London
gardens, the collector, ever on the alert for something
new, will naturally turn his attention towards the
country ;⸱ but a visit *en route* to the different sheets
of water which ornament the metropolis will be
found to be not altogether unproductive in the way
of specimens.

Chiefly conspicuous by their size are the large
swan mussels, *Anodonta cygnea* (Pl. III., fig. 3),
adult specimens of which will measure 6in. by 3in.,*
and the pearl-bearing *Unios* of somewhat smaller
dimensions. Not that there are any pearls now to

---

* A specimen of this mussel in the author's collection,
from the vicarage pond at Cowfold, Sussex, measures 7in. by
3½in. This is an unusually large one. But, some years
ago, several were taken out of a decoy pond in Firle Park,
Sussex, measuring 8 inches in length and 9 in circumference.
Cf. Merrifield, *Nat. Hist., Brighton*, p. 155 (1864).

1

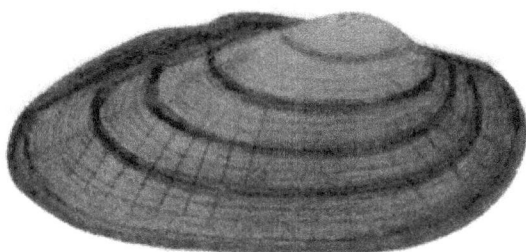

2

3

1. *Unio tumidus*. 2. *Unio pictorum*.
3. *Anodonta cygnea*. ½ nat. size. __

A. Rich. del.

W. West & C° imp

be found in London waters, whatever may have been the case formerly; but the collector may procure at least two species, *Unio tumidus* and *pictorum* (Pl. III., figs. 1, 2), which belong to the same genus as a pearl-bearing species, *U. margaritifer* (Pl. IV., fig. 3), formerly abundant in Great Britain, and still to be obtained in some parts of the country. In works which bear upon the subject, many rivers are noticed as having been at one time the seats of pearl fisheries; and, as every one knows, Britain nearly two thousand years ago was celebrated far and wide for its pearls :—so much so, indeed, that, according to trustworthy historians of that remote period, we are indebted to these precious jewels for the first hostile visit of the Romans to our shores, on which occasion, to use the words of a humorous modern writer, Cæsar broke in upon the *natives* with considerable energy. At the present day, we are informed that but one pearl on the average is found in every thirty shells; and, as only one in about ten is saleable, it requires the destruction of three hundred shells to find that one gem. At the present day, we are told, it is not unusual to find pearls in the Teith and Tay worth from £1 to £2 each.

But to return to the mussels of the London waters, *Unio tumidus* and *pictorum* (Pl. III., figs. 1, 2) are sometimes found in the same locality, and there are two or three varieties of each.

The former is a thick-shelled heavy species, oval in shape, black in colour, and with the hinge very prominent and swollen; hence its specific name. The latter is oblong rather than oval, broader in proportion to its size, and lighter; the shell thinner, of a green or olive-brown colour, finely striated, and the hinge less prominent. Both produce pearls, though of a very small and inferior size, and generally speaking the pearly secretion takes the form of an irregular mass deposited on the shell. The use which painters have found for the shells, viz., to hold their colours, has evidently suggested the specific name *pictorum*. Besides the London waters, including that in Battersea Park, the ponds at Hampstead and Highgate, the reservoirs at Kingsbury and Elstree, the river Brent between Hendon and Brentford, to say nothing of the Thames itself, are localities which may be named to such as are desirous of obtaining mussels; while those who delight in the exercise of a long country walk may give attention if

Plate IV

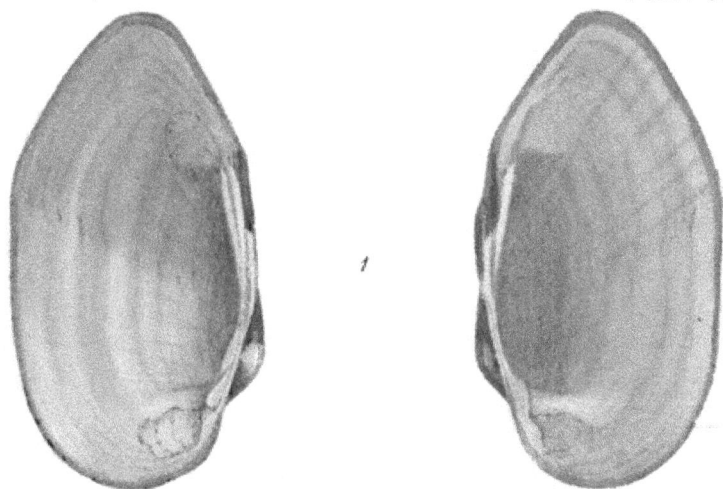

1. 2 Unio richensis. 3. Unio margaritifer.

A Rich lith.

W. West & C° imp

they please to the development of another kind of *muscle*, which is doubtless of greater importance.

When the water in the Regent's Park was drawn off after the lamentable skating accident in 1867, by which more than fifty persons lost their lives, a very pretty *Unio* was found partially scattered over the mud. From its shape and iridescent lining, in some specimens tinged with blue or lead colour, in others with pink or champagne colour, it was considered to be a new and undescribed species, and was provisionally named *Unio Richensis*, after a well-known collector who was the first to call attention to it at a meeting of the Linnean Society early in the following year. Although it has since been regarded by some conchologists as a mere variety of *U. tumidus*, we think a glance at our figures (Pl. IV., figs. 1, 2) will show the former opinion to have been not ill-grounded.

It is a little remarkable that, notwithstanding the depth of black mud in which these shells were found, not a single specimen showed any trace of erosion, although numberless examples of *Anodonta cygnea*, present in much greater abundance, were considerably eroded, as usual. Another circumstance noted was that, although the so-called *Unio Richensis* was highly iridescent in its lining, this was not the case with *Anodonta cygnea*. No specimens of *pictorum* were found there.

Mussels furnish food to many animals, but especially to otters, rooks, and crows. Numbers of these shells have been found in an otter's haunt, with the ends bitten off, and evident marks of teeth upon the broken fragments, the position of the shells indicating that the otter, after having crunched off one end, had sucked or scooped out the mollusc in much the same way as those who are partial to shrimps dispose of that esculent crustacean. Rooks and crows we have repeatedly observed in search of a dinner of mussels; and very systematically they set to work. On the muddy banks of the Thames at low water, and along the margin of the Brent, especially in time of drought, numbers of mussels may be found which have been opened by these birds. Should the shell occasionally prove a little too strong for them, they will fly up into the air with it and drop it from a height on hard ground, following it in the descent to find it broken, or to repeat the manœuvre until at length they get at the contents.

Occasionally they will wade a little way into shallow water in search of the mussels, if none are exposed on the bank, and it is amusing to watch them hurriedly take flight from the surface of the water as they now and then get suddenly out of their depth. Their efforts, also, to land a heavy mussel, and at the same time to save a ducking, are worth noticing.

Independently of its large size, the common swan-mussel, *Anodonta* (Pl. III., fig. 3), differs, as its name implies, from the *Unios* in the absence of teeth upon the hinge, although it would perhaps be more correct to speak of these processes as rudimentary only. Like the pearl-bearing mussels, this species produces eggs, which it retains within its shell, or, more correctly speaking, within its gills, until they are hatched; but the young are at first so unlike the adult that, as they have been frequently found adhering to the bodies of fishes, they have been erroneously regarded as parasites.

The large fleshy foot of these mussels enables them to travel considerable distances, and "ploughing the deep" may be said to be literally part of their occupation, as any one will admit who examines

the deep furrows which they make in the soft mud.
They feed on decomposed animal and vegetable
substances; and the size and solidity of the shell
depends on the abundance of the food and the
state of quiescence or motion, and of calcareous
matter in the water in which they happen to reside.
These shells make capital cream-skimmers, and in
French dairies are used for the purpose. They are
procured by means of a long-pointed stick, which is
thrust between the open shells when the animal is
feeding, and these, closing on the stick, allow it to
be drawn up out of the water.

Although a great many varieties of the common
swan-mussel have been described as different
species, we believe it is pretty generally admitted
that a second species of the genus (*A. anatina*)
exists in this country. It has a smaller shell,
which is longer in proportion, and the hinge line
is raised instead of being straight, while the posterior side is abruptly instead of gradually sloped
off. It is of similar habits, and frequents similar
situations to the last named.

The curious-looking zebra mussel (*Dreissena polymorpha*), although not indigenous to this country,

deserves a passing notice from the fact of its being now so generally distributed that it cannot fail to attract the attention of the collector.* It was first noticed in the Commercial Docks at Rotherhithe, and, being able to attach itself by a strong byssus to extraneous substances, there can be little doubt that it was introduced upon Baltic timber. It is now to be met with in canals and rivers in various parts of the country, and, through the New River, has even found its way into the streets of London. Some iron water-pipes which were taken up in Oxford Street were found to be in some places completely lined with these mussels, and the colour of the shells was as bright as if they had been always exposed to the light. In shape, the shell is oblong, with a sharp keel in the middle of each valve, and flattened below, with the end or beak pointed. The colour is a yellowish-brown, transversely barred on the upper part with darker brown, giving it that

* M. Marcel de Serres is of opinion that the *habitat* of *Dreissena polymorpha* was originally marine, from the circumstance of the shells being found in tertiary strata of marine formation. Pallas, by whom the species was first made known, described one variety of it as marine, and another as inhabiting fresh water.

striped appearance which no doubt suggested the
trivial name, " zebra " mussel (Pl. V., fig. 4).   It
is said to be met with occasionally even in unnavi-
gable waters (*cf.* Strickland, *Mag. Nat. Hist.*, N.S.,
1838, ii. p. 362 ; Bell, *Zoologist*, 1843, p. 253 ;
and Wolley, *Zoologist*, 1846, p. 1420).

Passing now to some of the smaller bivalves, the
collector will have no difficulty in procuring in and
around London two or three species of the genus
Sphærium, or, as it used to be called, *Cyclas*.   Let
the reader imagine little mussels somewhat spherical
or lens-shaped, and about the size of a sixpence or
less, and he will have some idea of their appearance.
They are found in canals, slow rivers, ponds, and
even ditches, and are very generally distributed.
The commonest kind, perhaps, is *corneum* (Pl. V.,
fig. 1), of which there are several varieties, and one
of which, *pisidioides*, was first described by Dr.
Gray (*Ann. Mag. Nat. Hist.*, xviii. p. 25) from specimens found in the Paddington Canal.   The shell
is nearly globular and equilateral, thin, glossy, and
of a yellowish horn colour, often with paler bands or
zones, denoting the periods of growth.   *S. rivicola*
(Pl. V., fig. 3) is much larger, and oval instead of

1. *Sphærium corneum* 2. *S. ovale*. 3. *S. rivicola*.
4. *Dreissena polymorpha*. 5. *Pisidium amnicum*
6 *Pisidium fontinale* 7. *Pisidium pusillum*.

A. Rich. lith   W. West & C.º imp

globular, with a more conspicuous ligament. The two are often found together, and well-known localities for them are the Grand Junction Canal at Paddington, the Thames shore at Battersea, Richmond, and Clifden Hampden, the ponds on Wandsworth and Clapham Commons, and the marshes below London. A third species, *Sphærium ovale* (Pl. V., fig. 2), has been found by Mr. Gwyn Jeffreys in the Paddington Canal, and, under the name of *S. pallidum*, it was figured and described by Dr. Gray (*Ann. Mag. Nat. Hist.*, xvii. p. 465) from specimens procured in the canal near Kensal Green. Some years previously, however, it had been met with in the Surrey Canal, but at the time it was supposed by the discoverer, Mr. Daniel, to be a variety of *S. rivicola*. It certainly resembles this species more than any other, but may be distinguished by its oblong and almost angular shape, thinner shell, and paler colour.

A nearly allied genus is *Pisidium*, of which there are five recognizable British species—*amnicum* and *fontinale*, with triangular-shaped shell; *pusillum*, oval; *nitidum*, round; and *roseum*, oblong—the three first named of which have all been met with

in various parts of the Thames, and marshes
around London (Pl. V., figs. 5, 6, 7). Formerly
these were all classed with *Sphærium* (or *Cyclas*, as
it used to be called); but, independently of their
smaller size, the species of the genus *Pisidium* differ
from those of *Sphærium* in the shape of their shells,
which are not equilateral—that is, the beak is situated
near the shorter end—and in having but one tube or
syphon instead of two. It may here be desirable to
explain that in the family *Sphæriidæ*, the mantle is
open in front, and forms at the posterior side a cylinder, which is often divided near its opening into two
contractile tubes, one for respiration and nutrition,
the other for excretion. The members of this
family, being gregarious, are often met with in
considerable quantities, the species of *Pisidium*
looking not unlike peas (hence their generic name)
scattered about. The distinguishing character of
the species in these two genera have puzzled many,
and some idea of the great variation which exists
may be formed from the fact that, out of forty-one
so-called European species of *Pisidium*, twenty-one,
according to Mr. Gwyn Jeffreys, are referable to
and are mere varieties of *P. fontinale*, and only six

species in all can be fairly recognized. Short of pointing out in what respects these six (or rather five, as British) differ *inter se*, we have, perhaps, said enough to stimulate research on the part of the collector.

The aquatic bivalves, then, may be thus enumerated—three *Unios*, two swan mussels, one zebra mussel, four species of *Sphærium* (or *Cyclas*), and five of *Pisidium*, or fifteen species in all. The aquatic univalves, or pond snails as they are often termed, are much more numerous, though the majority of them belong to two well-marked but very different genera—the flat coil-shells, *Planorbis*, and the thin-shelled, long-spired mud-shells, *Limnæa*—the remainder belonging to seven other genera, to be named presently. The various species of *Planorbis*, of which some six or seven are to be met with around London, frequent ponds, ditches, marshes, and stagnant water, and are generally found floating on the surface, or adhering to duck-weed and the leaves of other aquatic plants. Of the largest and most remarkable species, *P. corneus* (Pl. VI., fig. 4), we have already spoken (p. 16) when referring to the more noticeable generic forms

of land and fluviatile mollusca. Of the remaining
species, the flattened coil-shell, *P. complanatus;* the
keeled ditto, *P. carinatus;* and the white ditto,
P. albus, will be the most easily recognized. The
shell of *complanatus* (Pl. VI., fig. 3) may be distinguished from that of *carinatus* (Pl. VI., fig. 2)
by its narrower and more rounded whorls, as well
as by the keel being placed below, instead of in or
towards the middle of the periphery. It is usually
larger and thicker than that species, and is much
more plentiful, as well as more generally diffused.
The greyish white colour of *albus* (Pl. VI., fig. 1)
renders it sufficiently conspicuous, and on this
account it is not difficult to select it at once from
amongst a number of its congeners.

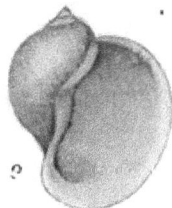

1. Planorbis albus 2. P. carinatus. 3. P. complanatus.
4. P. corneus. 5. Physa fontinalis 6. Valvata piscinalis.
7. Limnea palustris. 8. L. stagnalis. 9. L. auricularia.
10. Bythinia tentaculata

A. Rich. lith.   W West & C⁰ imp

CHAPTER IV.

In a little book entitled "Flora Metropolitana," published so long ago as 1836, and long since out of print, will be found an appendix in which the author, the late Mr. Daniel Cooper, gives a list of the land and freshwater shells found in the environs of London. Unfortunately, many of the names employed have become obsolete, or only hold good nowadays as synonyms, so that some little trouble is occasioned in identifying the species referred to; and, as several of the localities mentioned as the sites of ponds and marshes have been long since drained and covered with bricks and mortar, the collector might now search in vain for shells which formerly abounded there. Nevertheless, this list is instructive, as furnishing evidence of the former existence around London of species which may still be looked for in congenial haunts which have not been as yet interfered with. At the same time it carries with it a certain amount of authority; for amongst those who assisted the author in its compilation will be found the names of two well-known naturalists — Dr. J. E. Gray, whose edition of Turton's "Manual of British Shells" will always be a text-book for conchologists, and Mr. Thomas Bell, the well-known author of "British Quadrupeds," whose published researches in various branches of zoology can never be too highly estimated.

In this list we find ten species of *Planorbis* mentioned as occurring in the neighbourhood of London; but one of these is mentioned twice under different names, *marginatus* and *complanatus*; while *fontanus* and *imbricatus* of Cooper are respectively *nitidus* and *nautileus* of modern conchologists. The Hampstead ponds are referred to as the haunt of several species of *Planorbis*, such as *carinatus*, *vortex*, *contortus*, and *spirorbis*; and the Greenwich Marshes, and ditches about the Surrey Canal near Deptford, are said to have yielded *carinatus*, *marginatus*, *nautileus*, *corneus*, *contortus*, *albus*, and *spirorbis*.* Mr. A. F. Sheppard, in a list of

* *P. spirorbis* may often be found on grass in wet meadows.

shells found in the vicinity of Fulham, includes
Planorbis corneus, carinatus, and *vortex.* In ad-
dition to these localities, may be named the Brent
and the Lea, the reservoirs at Kingsbury and
Elstree, and ponds at Edgeware and Stanmore
Marsh, where most if not all of the above-named
species may be looked for with success. The mud-
shells (*Limnæa*) are quite as numerous and generally
distributed. Cooper, in the list referred to, gives
ten species as occurring in the neighbourhood of
London; but these are reducible to six, since two
of them, *scaturiginum* and *fragilis,* are respectively
the young and a variety of *stagnalis,* and a third,
glutinosa, a local species, belongs to the subgenus
*Amphipeplea.** The commonest are *peregra* (with
a variety *ovata*); *stagnalis* (Pl. VI., fig. 8), of which
there are many varieties; and *auricularia,* the ear-
shaped mud-shell (Pl. VI., fig. 9). The last named
has been met with in the ponds on Hampstead Heath.
Limnæa glabra has been met with in a pond near Nine
Elms, and formerly near Vauxhall, and *L. palustris*
(Pl. VI., fig. 7) is not uncommon in the marshes

* *L. glutinosa* has been found at Stanmore, Middlesex,
on the leaves of the yellow water-lily, *Nuphar lutea.*

E

below London. An interesting account of the habits
of *Limnæa stagnalis*, and its mode of respiration as
observed in confinement, was published some years
since by Mr. W. A. Lloyd, in "The Zoologist" for
1854, p. 4248. In similar, that is to say marshy
situations, are found two species of the genus *Physa*,
or bubble-shell, a peculiar genus intermediate between *Planorbis* and *Limnæa*. It resembles the
former in its long tentacles, and the latter in the
form of the shell, but has the spire sinistral. The
stream bubble-shell, *Ph. fontinalis* (Pl. VI., fig. 5),
may be found on watercress and other aquatic plants
in streams and canals, and is everywhere tolerably
common. The slender bubble-shell, *Ph. hypnorum*,
is rather more local, affecting ponds, ditches, and
rank grass in dried-up pools. Both are gregarious,
and may be recognized at once by the polished
appearance of their shells, the surface of which,
being more or less enveloped by an expansion of the
mantle, is kept bright by the lubricating friction
which it undergoes. The characters by which
fontinalis may be distinguished from *hypnorum* are
the oval instead of oblong shell, larger and wider
mouth, smaller number of whorls (that is, four or

five, instead of six or seven), shorter spire, and deeper suture. The foot of the animal in *fontinalis* is rounded in front instead of lanceolate, and the body is of a uniform greyish colour, instead of minutely speckled as in *hypnorum*. Dr. Gray considers these two generically distinct, and places the latter in the genus *Aplexus*, pointing out that, in *hypnorum*, the mantle has plain edges, and is not expanded over the shell, which has a long spire and an epidermis ; while in *fontinalis* the mantle is lobed, expanding over the shell, which has a short spire and no epidermis. But in regarding these differences as specific, and not generic, we have followed Turton, Forbes and Hanley, Gwyn Jeffreys, and other authorities.

Valvata piscinalis (Pl. VI., fig. 6) is a pretty little shell, which is not uncommon in ponds and still waters. The animal inhabiting it is remarkable for its branchial apparatus, which is external and resembles a plume, and for a curious appendage to the mantle to facilitate respiration, which looks like a third tentacle on the right side of the body. The mouth of the shell is closed with an epiphragm or valve ; hence the generic name, while its partiality

E 2

for fish-ponds no doubt suggested the specific name *piscinalis*.

Another little water snail is *Bythinia tentaculata*, which in appearance (Pl. VI., fig. 10) is not unlike a miniature *Paludina*, already described; but in the former genus the animal is oviparous instead of ovoviviparous, and sessile-eyed instead of stalk-eyed. Mr. Gwyn Jeffreys has pointed out that, although the derivation of the word *Bythinia* would imply that these molluscs inhabit deeper water than others of the same family, such is not the case. They generally frequent small streams, canals, shallow ponds and ditches, especially in the marshes below London, where they lay their eggs in three long rows on stones, as well as on the stalks and leaves of water plants. The animal floats or creeps on the under surface of the water, and is said to be carnivorous as well as herbivorous. It has been found commonly in ditches at Fulham. A second species of the genus, named *Leachii*, after the late Dr. Leach, has been found in the Woolwich Marshes, but it is much more local than the last named, and less abundant. In company with the two last named may be found *Hydrobia similis* and *Assiminia grayana*, but these are, more correctly speaking, brackish-water shells. *Hydrobia similis* resembles *Bythinia leachii*, but may be distinguished by its smaller size and grooved suture; the operculum is horny, concentric, and the nucleus lateral; whereas in *Bythinia* it is somewhat shelly, and marked by concentric ridges having the nucleus central. Mr. Gwyn Jeffreys states that this species is found in muddy ditches, occasionally overflowed by the tide, by the side of the Thames from Greenwich to below Woolwich. These ditches are separated from the river by a high and broad embankment, which is provided at distant intervals with sluices to drain off the surface water. It lives there in company with *Bythinia tentaculata* and other fresh-water shells, as well as with the more marine and peculiar mollusk *Assiminia grayana;* and it is gregarious. Its food appears to consist of decaying vegetable matter; and its habits are rather active, creeping and floating with tolerable rapidity. Mr. Prestwich and Mr. Pickering found specimens of it in peat, in the main drainage cutting between Woolwich Arsenal and the exit to the Thames, through Plumstead Marshes; but it can scarcely be considered one of our upper tertiary fossils.

Assiminia grayana differs from *Hydrobia* in not having the eyes placed on tubercles, and from the marine *Rissoa* in the tentacles being united to the eye-stalks, which equal them in length. The shell, of a liver-brown colour, is ovate-acute, with five whorls, and about a quarter of an inch in length. The suture is slightly impressed; there is no umbilicus; the aperture is ovate; the operculum horny, ovate, and of a blackish-brown colour. It inhabits the banks of the Thames between Greenwich and Gravesend, and is tolerably abundant, living on the mud beneath the shade afforded by *Scirpus maritimus* and *Festuca arundinacea.**

---

* "The number of estuarine species," says Mr. Tate, " which have a place in our works devoted to British land and freshwater snails is very few, and the majority, moreover, are confined to the margins of the tidal rivers in the South of England. Thus *Assiminia grayana*, *Hydrobia ventrosa*, and *H. similis*, live on the mud-banks beneath the shade of sedges and rushes, skirting the Thames below Greenwich. To gather these small shells singly is a tedious operation; but if a thin piece of flat wood, or other substitute as the ingenuity of the collector suggests, be used to scrape lightly over the surface of mud, transferring the mass to the *dredger*, or tin sieve and washing in water, a number of specimens, sufficient to stock every private cabinet in the country, may be obtained in a short space of time."

3

4

5

6.

7.

8.

1. *Ancylus fluviatilis* 2. *A. lacustris* 3. *Neritina fluviatilis*
4. *Paludina vivipara*. 5. *P. contecta*, 6. *Vitrina pellucida*
7. *Succinea putris* 8. *S. elegans*.

A. Rich. lith.

W West & Cº imp

The two species of *Paludina*, which, as above stated, are closely allied to *Bythinia*, are *vivipara* and *contecta* (Pl. VII., figs. 4, 5). They are pretty generally dispersed, inhabiting rivers, canals, and large sheets of water, and their distinguishing characters have already been pointed out (p. 13). In the timber docks on the Thames, and in the various canals about London, these shells are very abundant. Specimens from a pond on Hampstead Heath were found to have their apices much eroded, which was due, no doubt, to the action of sulphuretted hydrogen given off from decomposing animal and vegetable matter. The variety *unicolor*, without bands, has been obtained by Mr. Gwyn Jeffreys in the Thames at Richmond.

In waters which have a stony or gravelly bottom may be found a pretty little shell about the size of a pea, with a very short spire and semilunar mouth, solid and glossy, and of a purple colour spotted with white. This is *Neritina fluviatilis* (Pl. VII., fig. 3), the only fresh-water species of the genus to be found in this country, although there are several marine forms which are also met with in brackish water. It may sometimes be found on the

submerged leaves of *Nuphar lutea*. It is generally
encrusted with mud, the removal of which appears
to destroy the beautiful coloured markings which
are so ornamental when the shell is found dead
and empty.

The so-called freshwater limpets, *Ancylus fluviatilis* and *lacustris*, or *oblongus* as it is often called
(Pl. VII., figs. 1, 2), furnish another illustration of
the fact that both salt. and fresh waters have their
respective representative forms. Both species may
be found in the Thames and the Lea, adhering to
stones and leaves of aquatic plants, especially on the
leaves of *Nuphar lutea*, where they furnish abundant
food for fish, which push them off with their snouts
and swallow them—in much the same way, no doubt,
as the Oyster-catcher (*Hæmatopus ostralegus*) over-
turns the marine limpet (*Patella*) with his bill to get
at the animal within. The freshwater limpets are
frequently sought after as food by the Dipper, Moor-
hen, Water-rail, and different species of Grebe, as evi-
denced by the fragments of shells which have been
found in the stomachs of these birds. They also swallow the bubble-shells (*Physa*) and *Valvata piscinalis;*
but, although the Little Grebe or Dabchick may be

observed, sportively as it were, picking off these tiny molluscs from the weeds in our streams and mill-ponds, we believe that the staple article of food with all the Grebe family is fish.

Although a few other aquatic species may be discovered by the industrious conchologist, we believe we have now mentioned all that are most likely to be met with by collectors in the environs of London.

There are still a few terrestrial mollusca which, unlike those already named, are unable to endure the smoky atmosphere of town and suburbs, and must accordingly be sought for in the purer air and more congenial haunts of wood and field. Some, as *Helix virgata, caperata, cartusiana,* and *ericetorum,* although deserving of a place in a list of shells found in the environs of London, are nevertheless peculiar to the chalk, and not to the London clay, so that it may be convenient perhaps to defer noticing them for the present, until we have concluded our remarks on those which are not restricted to a chalky soil, and which may therefore be met with much nearer to the metropolis. Moreover, by bearing in mind this distinction of habitat, the memory will be assisted in discriminating many allied but distinct species.

The amber snails (*Succinea putris* and *elegans*) in form and habits resemble the mud snails (*Limnæa*), as also in some respects the true snail (*Helix*), being to a great extent amphibious. Sometimes they may be seen crawling on stones under water, on mud, or on the leaves of various aquatic plants; at other times they may be met with in comparatively dry spots at a distance from water. The Reed Bunting, Bearded Titmouse, and other small birds, which are fond of feeding by the water's edge, take quantities of these tiny molluscs. In the stomach of the Bearded Titmouse has also been found *Pupa muscorum*. The amber snails are very sluggish in their habits, and secrete a quantity of slime. The shells in appearance resemble those of *Limnæa*, but are thinner, more transparent, of an amber colour —hence the popular name for them—and have no fold on the columella or pillar (Pl. VII., figs. 7, 8). The two species named above have been considered by some conchologists to be mere varieties of the same species, great variability of form being observable in all the species of the genus *Succinea*; but *elegans*, or *Pfeifferi*, which is another name for it, is said to differ from *putris* in the darker colour of its body and the more slender shape of the shell, as well as in its longer and more pointed spire.

Dr. Gray, in his edition of "Turton's Manual" (pp. 146, 147), treats them as varieties of one species. Capt. Bruce Hutton, of the 61st Regiment, writing in "The Zoologist" for 1862 (p. 8138), records the following observations, which lead him to believe they are distinct. He says: "Between the North and South Camps, Aldershott, runs the Basingstoke Canal, along the sides of which, both up and down as far as I have been, the *Succinea* abounds; and they are all alike, small, narrow, very oblique, and, while the animal is in the shell, the colour is bluish-black. About a mile from Ash the Canal is raised on an embankment nearly thirty feet above the level of the surrounding country, and the land at the foot of the bank is used as an osiery, where among the willows, etc., the yellow Iris grows luxuriantly. Last month (May) nearly every leaf in this spot had a *Succinea* on it, some of the largest I have ever met with, and about four times the size of their neighbours on the Canal. The colour of the animal was invariably a dull white, or white with a shade of yellow. The shell much larger, the whorls more

convex, the suture not so oblique, and the mouth
broader in proportion; it has also a varnished appearance, which is wanting in the smaller kind. I kept
many of both kinds alive for a considerable time, and
watched their habits. Several paired in captivity
and deposited eggs, but though kept together they
never united, except with their own sort. The large
kind I observed rather to avoid the water, whereas
the small often took to it of their own accord, and
remained in for a length of time, particularly at
night. I have this year lost dozens of the smaller
Succinea through the ravages of a small worm, probably a *Cochleoctonus*. More than once, while looking at my captives, I have noticed an individual
become restless and begin to throw its head from
side to side. Shortly after a worm has made its
appearance, usually by eating its way through the
right side of the neck of the *Succinea*, just above and
behind the genital orifice. The poor victim seldom
lived more than an hour or two after the exodus, and
seemed to die in great pain, as the genital organs
and 'poche buccale' often protruded, as if they had
been squeezed out by strong convulsions." The
writer adds, by way of postscript (p. 8171):—

"After dissecting and examining many specimens under a microscope, I have found that the upper jaw is in *S. putris* divided into three teeth, one large projection in the centre and two small ones, one on each side; whereas, in *S. pfeifferi*, there is only one tooth-like projection in the centre of the jaw. I have found this to be invariable in very many specimens of both kinds that I have examined; and the fact goes far to convince me that *S. putris* and *S. pfeifferi* are more distantly connected than some conchologists seem to think."

Mr. Gwyn Jeffreys, who considers *elegans* or *pfeifferi* distinct from *putris*, has met with specimens of the former in the neighbourhood of Hammersmith. A third British species, *oblonga*, is much rarer, or more correctly speaking more local, being generally met with in ditches near the sea coast. We know of no instance of its occurrence near London, although Cooper, in the list to which we have already referred, includes *oblonga* as found by Mr. James Carter at Hammersmith. We suspect that he means *elegans*, which is more oblong in shape than *putris*, the only other species which he included, although by another name, viz., *amphibia*.

Only one species of glass snail (*Vitrina*) inhabits this country; although a second formerly did so, as proved by its occurrence in our upper tertiary strata. *Vitrina pellucida* (Pl. VII., fig. 6) is not uncommon under dead leaves, moss, and roots in woods and shady places, and seems partial to moisture, being always more active after rain. The shell is very thin, brittle, glossy, and transparent; the spire is short and blunt, the suture very slight, and the mouth nearly oval; in shape not unlike the shell of *Zonites*. The best time to look for this species is in autumn; but care should be taken in turning over dead leaves, moss, &c., in search of it, since from its extreme brittleness the shell is very easily destroyed. It would seem that the animal inhabiting it is a favourite morsel with the Hedgehog; numerous fragments of shells having been found from time to time in stomachs examined. This is also the case with some species of *Zonites*.

Where the soil is calcareous, search may be made for the needle agate shell, *Achatina acicula* (Pl. VIII., fig. 2), which is often found some inches below the surface. To give some idea of its appearance, it may be observed that specimens before now have been

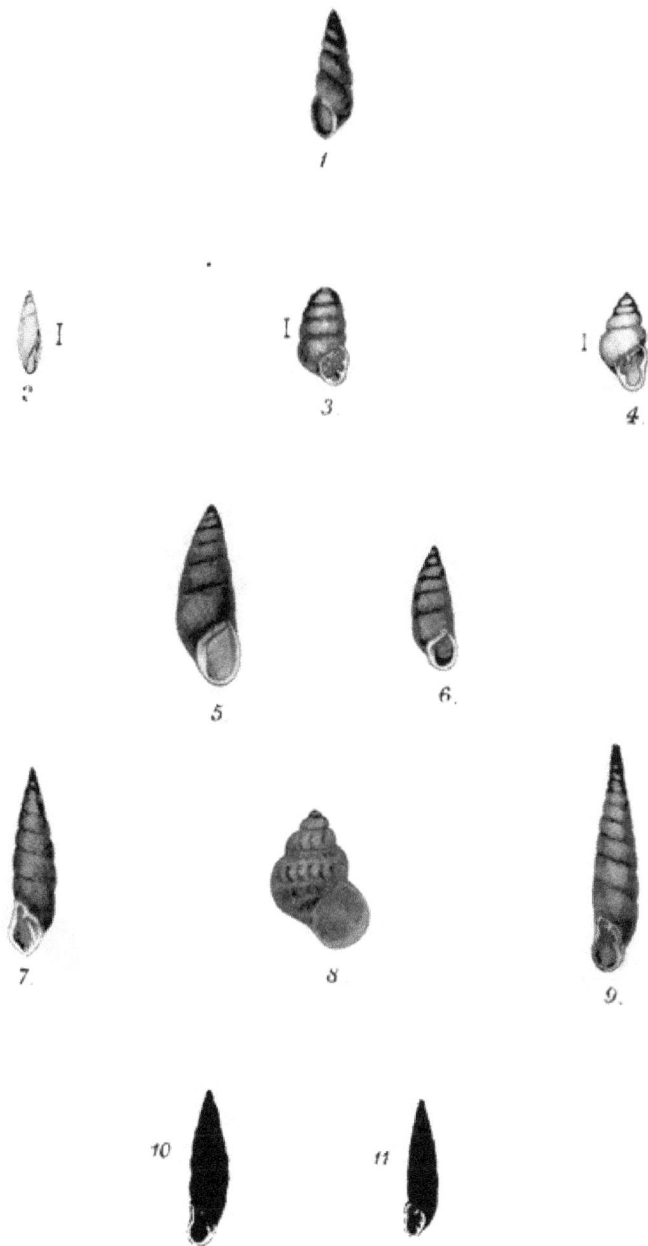

1. Balia perversa 2. Achatina acicula 3. Vertigo pygmaa.
4. Carychium minimum 5. Bulimus montanus. 6. B. obscurus
7. Clausilia biblicata 8. Cyclostoma elegans 9 Clausilia lamunata.
10. C. rolphii . 11. C. rugosa.

W. West & Co imp

mistaken by the uninitiated for little white maggots, the shell being long, thin, and cylindrical, and rather smooth. The Swedish naturalist, Nilsson, who has given a good description of this mollusk, observed that through the transparency of the shell the irregular motion of its breathing could be easily seen with a glass, and that when the respiratory cavity was shut the motion ceased, but was continued when the chamber was reopened; and he compared this alternate expansion and contraction of the breathing organ in this snail to the pulmonary action of vertebrate animals. He supposed that it fed on the tender and jucy fibrils of the roots of grass. The little sedge shell, *Carychium minimum*, may be looked for at the roots of grass and other plants, and amongst moss in damp situations. Specimens have been found on the roots of the *Iris pseudacorus* at Hammersmith, and in winter it may be discovered in the hollow stems of the larger umbelliferous marsh plants. From its very minute size, being one of the smallest of our land shells, a very careful search for it is necessary. In appearance the shell is not very unlike *Pupa*, already described, but is more transparent, with fewer whorls, and has the spire less blunted (Pl. VIII., fig. 4).

The fragile moss-shell (*Balia perversa*, or *fragilis*), as its name implies, may be sought for upon moss-grown trees and rocks, and is not uncommon. Cooper mentions it in his list as found at Hammersmith at the foot of trees, concealed by grass; but since this list was published, trees at Hammersmith have given way to such an extent to bricks and mortar that this locality must now be accepted with reservation. In "The Zoologist" for 1851, p. 3121, Mr. A. F. Sheppard includes it in a list of shells found in the neighbourhood of Fulham, as having been met with on old walls. Although a small shell, its shape more nearly resembles *Clausilia;* that is, instead of being somewhat cylindrical throughout its length, it has a wider mouth, and tapers gradually throughout the length of the spire. The spire is sinistral or reversed, as in *Clausilia* (Pl. VIII., fig. 1).

Bulimus obscurus and *montanus* (Pl. VIII., figs. 5, 6) may both be found adhering to the trunks of trees, especially beech,* ash, and hornbeam, and at

---

* The beech is an especial favourite with snails, more so than any other tree, and after showers the trunks may be seen studded with them. The probable explanation, says

first sight, and until the eye gets accustomed to their appearance, might well be mistaken for chrysalides, or little excrescences of the bark. The Rev. Revett Sheppard, writing of Suffolk mollusca, observes that " these shells, particularly in their young state, show great sagacity and ingenuity, by covering themselves with an epidermis adapted to the different situations in which they are found ; and, when so covered, it is almost impossible for any other than a conchological eye to detect them. If its abode be upon a tree covered with lichens, then is the epidermis so constructed as to cause the shell to resemble a little knot on the bark covered with such substances. If on a smooth tree, from whose bark issue small sessile buds, as is frequently the case, it will pass very well for one of them; and on a dry bank, or the lower part of the body of a tree splashed with mud, its appearance will be that of a little, misshapen, pointed piece of dirt."

Bulimus montanus prefers a southern aspect, and ascends trees to a height which frequently renders it

Mr. Tate, is, that this tree is resorted to by the snails for the purpose of feeding upon the minute parasitic vegetation with which the trunk is clothed.

F

undistinguishable. The supposed security of these
snails and some others, in localities where in fact
they are most abundant, is doubtless due to this
peculiar habit of ascending trees during the summer
months; for at this season only dead shells will
reward a search amongst the herbage at the
foot.

In similar situations are found the close shells
(*Clausilia*), of which three out of four British
species have been met with in the environs of
London.

These shells are worth examining carefully, on
account of a remarkable peculiarity in structure,
namely, a lid to the mouth of the shell, which closes,
so to say, with a spring. If the outer part of the
last whorl be broken off, there will be found a spoon-
shaped calcareous plate, or valve, attached to the
column of the spire by an elastic filament. When
the animal protrudes from its shell, this plate is
thrust aside, and when it withdraws it closes the
door, as it were, behind it.

Clausilia biplicata (Pl. VIII., fig. 7) is not uncommon on the banks of the Thames near London;
and specimens have been obtained in Hyde Park

near the Serpentine. *C. rugosa* (Pl. VIII., fig. 11) has been found by Mr. Gwyn Jeffreys on marshy ground near Battersea; and *C. laminata* (Pl. VIII., fig. 9), although not so common, is nevertheless reported an inhabitant of trees upon the London clay. The fourth British species, *C. rolphii* (Pl. VIII., fig. 10), is very local; but we shall have occasion to notice it at greater length when treating of the shells which inhabit the chalk.

We have not pointed out the distinguishing characters of the different species of *Clausilia*, and it is by no means easy to do so in a few words. Some idea of their appearance, size, and colour may be gained by reference to the figures on Plate VIII., but in this, as in other cases, the collector will derive the best assistance from an inspection and comparison of well-authenticated specimens in the cabinet of some friend. We may direct attention, nevertheless, to the variation in the form of the mouth in the different species, and to the number and arrangement of the so-called teeth.

CHAPTER V.

So far as we are aware, there are no aquatic
mollusca, either bivalves or univalves, peculiar to
the chalk, and almost all that have been mentioned,
as inhabiting the London clay, are equally at home
on the lighter and whiter soil. But amongst the
terrestrial univalves there are several which are
only to be found upon chalky ground, and must
therefore be sought at a distance from the metropolis. The greater number of these belong to the
true snails (*Helix*), most of which are well marked
and easily recognizable species.

With regard to the largest of them, the edible
snail, *Helix pomatia* (Pl. IX., fig. 4), there
appears to be some doubt whether it is an indigenous or imported species in this country.
Some have supposed it to have been introduced
by the Romans; but there seems to be no other
foundation for this idea than that it is found in the

Plate IX.

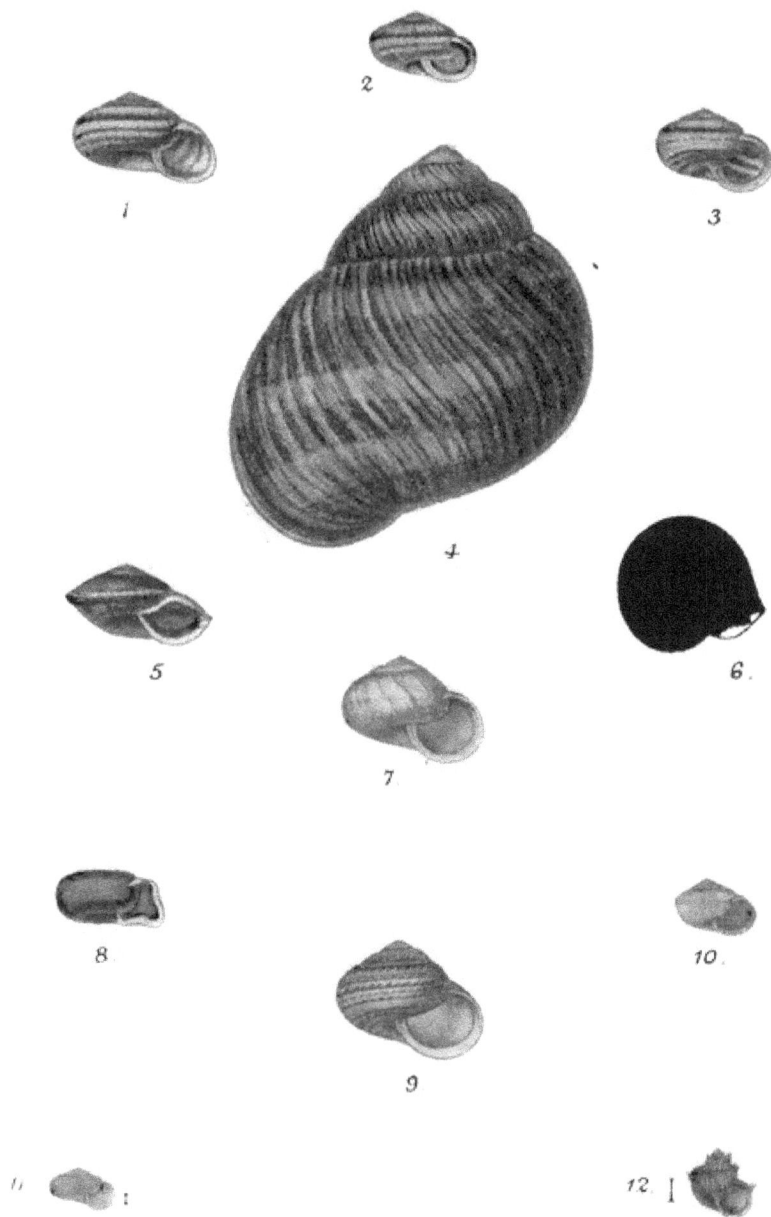

1. Helix ericetorum. 2 H. caperata. 3. H. virgata. 4. H. pomatia.
5. 6. H. lapicida. 7. H. cantiana 8 H. obvoluta. 9. H. pisana. 10. H. carthusias
11. H. pygmœa. 12. H. aculeata.

vicinity of many ancient encampments. Several
such sites, moreover, could be named where this
large snail is not found; and, as in Central Italy its
place is supplied by another species, *Helix lucorum*,
it is not by any means certain that it was known to
the Romans.

Its large size—the mouth alone measuring an
inch in diameter—and the consistency of its body,
will readily distinguish it from all others of the
genus to which it belongs. It is furnished with a
peculiar epiphragm, or mouth-piece, by means of
which it effectually closes the entrance of its shell,
and keeps out the cold and wet while it is hybernat-
ing. This lid, as we may term it, exactly fits the
mouth of the shell. It is a solid calcareous plate,
slightly convex, and is secreted and formed by the
mantle, resembling, until it hardens, liquid plaster
of Paris. After the animal has hybernated and
the fine weather has returned, this covering is
no longer needed, and is cast aside. A French
naturalist, M. Gaspard, who has paid considerable
attention to the structure and economy of this
species, says that when the period of hybernation
arrives these snails get indolent, lose their appe-

tite, and become gregarious. Each snail then, with its large and muscular foot, scoops out a hole in the ground, just large enough to contain its shell. This it roofs in, and lines with earth and dead leaves, and after making with its slime a kind of mortar, it uses it to smooth over the inside of this winter domicile. It then closes the mouth of the shell with the lid already described, and withdraws as far as possible into the interior of its shell, covering as it retires the empty spaces with several layers in succession of a fine membrane or film, in order more completely to exclude the air. In this snug receptacle it remains in a torpid state until the return of spring, all animal functions being in the meantime suspended. It then loosens and casts aside its winter bonds, and resumes its former life.

In May these snails propagate, and in June lay their eggs, producing generally but a single brood in the year. The eggs are about the size of a small pea, and in colour and consistency somewhat resemble the berries of the mistletoe. They are deposited in a sort of nest made in the loose earth, which protects them from wet and too much heat.

No incubation takes place, but they are left to the care of nature; and in three or four weeks, according to the state of the weather and temperature, the young are duly developed.

On the chalk hills about Dorking and Reigate, Mickleham, Boxhill, and in some parts of Kent, this large snail is tolerably common, but we have looked for it in vain upon the South Downs, where we should certainly have expected to find it. Mr. W. Jeffery of Ratham, near Chichester, writing to us so recently as January 1874, says, "Many times have I searched for *Helix pomatia* on our downs, but always without success. Some five or six years ago I had between thirty and forty sent me from the Surrey Downs, a part of which I turned down in my garden, and the remainder on a bank of light soil near. Of those on the bank I saw no more, but those in the garden seemed to do pretty well for a time, and at least one brood of young were hatched, some of which attained the full size. Now, however the old stock is no more, and last summer I saw only two of those bred in the garden. These are of of a much darker colour than the imported speci- mens, and in their earlier stages of growth led me to think they were hybrid with *H. aspersa.* In the garden *H. pomatia* is not nearly so destructive as *H. aspersa,* preferring as a rule decaying vegetation ; a yellow, half rotten, and then glutinous turnip leaf is particularly a favourite morsel with them."

In a copse upon the downs in West Sussex, not far from Petersfield, one or two dead shells have been found, from which it may be assumed that *Helix pomatia* formerly existed in that neighbourhood, but no living examples have been met with in recent times. An enterprising friend, however, with a taste for acclimatisation, turned out in that locality fifty or sixty live specimens, which were procured at Preston Candover, where thirty years ago they were plentiful ; but the experiment to establish them in the new locality failed, for not one of them was seen afterwards ; and if, as we suspect, the hedgehog found them toothsome, the fact speaks volumes for the discriminating palate of our erinaceous friend, who in this instance only emulated certain epicures amongst our Gallic neighbours. With the latter a dish of escargots is a recognized dainty, and in the summer months may be obtained at the restaurants of most of the principal towns in France as readily as whitebait is procured at Greenwich. Some idea may be formed of the estimation in which these snails are held there, from the fact that Burgundy and Champagne alone send no less than 100,000 of them daily to Paris. In *The Field* of April 19, 1873, appeared an article on the culture of edible snails in France, in which some very interesting details are given; and the so-called apple snail, *H. pomatia*,* is of course especially noticed. To this account we may refer such of our readers as have not already perused it.

A curious circumstance is related in Merrifield's "Sketch of the Natural History of Brighton," p. 157 (1864), which proves that rats are as fond of snails as some members of the human race, and are quite as ingenious in capturing them. The facts were thus narrated to the author by Mr. W. W. Attree of the Queen's Park:—"While my father was building this house (the villa in Queen's Park) the gardens, laid out beforehand, were colonized on a

* With regard to the name "apple snail," which is applied to this species, it may be appropriate as regards its shape, or with reference to the animal's *penchant* for apples; but the word "pomatia" is derived from πωμα, an operculum, and not from "pomum," an apple.

sudden by crowds of rats. That they should travel
half a mile from the town was not strange ; but there
was no inhabitant near the unfinished walls, and
apparently nothing more to tempt their visit than
when the spot was a bare hill-side. The workmen
said that the rats came for the new plastering;
but that, if possibly a *bonne bouche* in rats' diet,
could not, it seemed to me, support them. Besides,
they could scarcely have eaten it without their
depredations being discovered by the workmen, and
this did not take place. While still wondering
about the matter, I one day watched a rat come
out of his hole at the foot of a mound in the back
garden, go some paces without perceiving me, climb
the stalk of a hollyhock, clear off several snails,
bring them down in one paw, like an armful, and
run with them on three legs into his hole.   On
examining this hole and others as well, I found the
inside strewed for some distance with broken snail
shells.   At that time there was about the place a
great variety of snails with delicately coloured shells
of different sorts.   I fancy they have been cleared
off by the pea-fowls who regularly hunt the ground,
the pea-hen quartering the ground like a pointer.

We have hardly any, except the common brown snail, now left. I looked among the *débris* round the rats' hole, to see if they had chosen any particular kind of dainty snail, as the Romans did, and some moderns have done, but the broken shells were almost all those of the common brown snail, with only a coloured one here and there among them."

Thickly sprinkled over the chalk hills and sheep walks of the South Downs may be found the heath snail, *H. ericetorum*; the zoned snail, *H. virgata*; and the wrinkled snail, *H. caperata* (Pl. IX., figs. 1, 2, and 3). After rain, numbers of these may be seen scattered over the downs, clinging to the grass stems or leaves of the different shrubs with which the downs are studded; and, from the fact of their being gregarious and so abundant, the popular notion has arisen that it sometimes rains snails. Mr. Gwyn Jeffreys thinks that the idea of their descending in showers may also have originated in a whirlwind having caught up a number of them, by sweeping along a grassy plain, and dropping its contents in a limited area. Borlase, in his "Natural History of Cornwall," speaks of these snails as " yielding a most

fattening nourishment to the sheep" which feed upon
the downs, and pick them up with the short grass; and
a similar observation has been made in other parts of
the south of England, notably at Dartmoor and on
the Hampshire and Sussex Downs. We have often
wondered how these molluscs contrive to withstand
the glaring heat of the sun upon the exposed downs,
especially in the position in which they are often
found, that is, sealed to a grass-stem, a foot or more
from the ground.

The flat shape of *ericetorum*, its usually large
umbilicus and nearly circular mouth, will readily
serve to distinguish it from any other of the banded
snails. In Sussex the village children collect them
by bushels, and, threading them on string, make
necklaces and bracelets of them. Mr. W. Jeffery
informs us that *ericetorum* attains a larger size in
the valleys on the north side of the downs, where
the herbage is less scanty than on the south side
and the sun has less power. The second species,
virgata, is of a white or cream colour, with a single
broad band of purple-brown just above the periphery,
and several narrower bands below it. In this respect
it differs from *pisana* (a more local species found

nearer the coast),* which has only one band on the
body whorl (Pl. IX., fig. 9).

The colour of these shells, however, varies very
much, being occasionally plain yellowish, white, or
dark brown with white bands, or the dark bands are
streaked or interrupted so as to make the surface
appear spotted. *Helix caperata* differs from *virgata*
in its much smaller size, depressed spire, and
larger umbilicus, and especially in the numerous
rib-like striæ which hoop round each whorl. *Virgata*
may be found as near London as Lewisham and
Charlton, amongst thistles, nettles, and other rank
herbage about chalk pits; and around Woolwich
it was formerly not uncommon. It is said to feed
on Lady-birds (*Coccinella*) and other small insects.
Caperata is also found around Lewisham, and is
plentiful at Boxhill. The heath snail, *ericetorum*,
has been met with as near London as Charlton,
Banstead Downs, and the roadside between Dartford
Heath and Green-street Green. A bank between the
fifteenth and sixteenth milestones on the Sevenoaks-
road has also been recorded as a locality for this snail.

* The specific name "Pisana," was given to this shell,
from its having been first met with at Pisa.

All three species are plentiful in the neighbourhood of Reigate, where *caperata* is especially common on palings and nettles. On the stems of beech trees, and often concealed by ivy, may be found the rock snail, *Helix lapicida* (Pl. IX., figs. 5, 6). It is about the size of *ericetorum*, but much more solid, lens-shaped, of a dark brown colour, and with a sharp edge or keel round it, which distinguishes it at once from all others of its kind. The crevices of rocks and old walls are favourite situations for this snail; but, as it is rather inactive by day, the best time to look for it is at twilight, or after a shower of rain. The inappropriate name *lapicida* was bestowed upon it by Linnæus under the erroneous impression that it bored or excavated calcareous rock, as the Teredo does wood.* This is one of the very few in-

* Perforations on the under surface of projecting limestone crags, which *H. aspersa* and some other species are found occupying as winter quarters, have been regarded as the result of a constant resort for shelter to the same spot, winter after winter. The erosion is believed to be due to the action of the foot, aided by an acid secretion; although another theory is, that the snail works with its shell after the fashion of an auger. It seems not improbable, however, that the snails abrade the walls of these limestone cells with their tongues, for the purpose of obtaining the carbonate of lime.

stances in which a species has been inaptly named by that most remarkable and observant of naturalists.

We have found this snail in Gatton Park near Reigate, and on Reigate Hill, and have obtained numerous specimens in the neighbourhood of Chichester, and in the beautiful valley known as Kingley Vale. The Kentish snail, *H. cantiana* (Pl. IX., fig. 7), is by no means rare, inhabiting hedges, wooded banks, and walls in the home and many of the southern counties of England.* The shell is

---

* Referring to the distribution of *Helix cantiana*, which is generally supposed to be restricted to the south of England, Mr. John Hawkins writes in *The Field*, of 24th January, 1874, as follows:—"Two years since, when taking a friend to inspect the Roman camp at Honington, while searching for shards of old pottery, I found *Helix cantiana*. As it was in the winter season, only dead shells were procurable. Upon my next visit, I was fortunate enough to discover two or three live specimens. I could hardly believe, at first, that the Kent snail should occur in Lincolnshire, and was inclined to attribute the finding it to any cause but the true one. However, descending into the valley, and hunting in a dyke for some geological specimens among the stony *débris*, I found a whole colony of these *Helices*, and on comparing them with specimens brought from the Undercliff, Isle of Wight, I found them quite the equals of these in every respect. Since that time I have

not nearly so flat as the four last-named species,
but yet is not so globular as that of the common
garden snail, *aspersa*, or the apple snail, *pomatia*,
being more compressed above and below. It is
almost of a uniform white or fawn colour, although
sometimes marked with a light but indistinct spiral
band, which is placed a little above the periphery,
and does not extend much beyond the last half of
the body whorl. In the same localities mentioned
for *lapicida* we have found the Kentish snail
not uncommon, and this is especially the case
in the neighbourhood of Reigate. A much smaller
mollusc is *Helix cartusiana* (Pl. IX., fig. 10),
so called from having been first discovered near
a Carthusian monastery. It has a more solid
and nearly opaque shell, that is, much less glossy
and transparent than *cantiana*, of a light brown or
fawn colour, generally encircled with a whitish spiral
band, placed a little above the periphery. It is
generally found attached to grass stems and weeds in
the hollows of the downs, and is tolerably common.

procured them on the line of the old Roman road which
intersects our heath district, and there is no doubt that they
occur all along the stony district of Lincolnshire."

We have heard of its being taken about chalk pits at
Lewisham and Charlton, and even on Hampstead
Heath. It is common on the South Downs, near
Lewes, and about Beachy Head, where it is found on
the short herbage clothing the chalk-downs.

Beech woods are very favourable to many of the
species just named, and, for some reason or other,
shells procured in such situations are generally larger
and better coloured than those found in more exposed
haunts. Doubtless, shade and moisture are indis-
pensable to the growth and healthy condition of the
animals inhabiting them.

The cheese snail (*Helix obvoluta*), so called from
the cheese-like shape of the shell (Pl. IX., fig. 8),
is so local a species that it is supposed to have been
accidentally introduced at no very distant date. As
an inhabitant of Ditcham Wood, near Buriton, it
was first discovered by Dr. Lindsay, at one time a
resident in the neighbourhood. It has since been
found at Ashford Wood and Stonor Hill (Rev. W.
H. Hawker), Uppark, near Petersfield (Mr. J.
Weaver), Kingley Vale, near Chichester (Mr. W.
Jeffery, Jun.), and Crabbe Wood, near Winchester,
(Mr. W. A. Forbes), amongst moss at the roots of

hazel, and on beech trees at some height from the
ground ; but these are the only localities for the
species in England with which we are acquainted.

It is so very unlike any other British *Helix* that
it cannot fail to be recognized at once. The shell,
as we have said, is cheese-shaped—but it may be
desirable to add of the shape of a Cheshire cheese,
some cheeses, as the Dutch, being globular. It is
solid and opaque, of a reddish-brown colour, the
eperdimis clothed with stiff hairs, and with the outer
tip rose-coloured, very thick and reflected. The
umbilicus is large, exposing part of the whorls, and
all the internal spire. During the period of hybernation the mouth of the shell is closed with a thick
chalky epiphragm, which contrasts strongly with the
rich reddish-brown colour of the shell itself.

In similar situations—that is, among dead leaves
and moss in woods, as well as under fragments of
chalk — may be found the little prickly snail, *Helix
aculeata* (Pl. IX., fig. 12), the shell of which measures about the tenth of an inch in breadth, and the
same in height. Its distinguishing character is that
the epidermis with which it is clothed rises, at frequent and regular intervals in the middle of each

whorl, into sharp teeth or points, so as to present under a lens the appearance of a very elegant spiral of bristles. A friend, who has paid some attention to the land shells in Sussex, affirms that he has found pieces of decaying bark an excellent lure for many of the smaller snails, and especially the one just named, *aculeata*. Mr. A. F. Sheppard has met with *aculeata* on the trunks of oak trees at Fulham (*cf.* "Zoologist," 1851, p. 3121).

We should include amongst the shells of the chalk district, *Helix rotundata*, had we not already referred to it (p. 27) as being found upon the London clay. With the mention of *oculeata*, therefore, we may bring our remarks on the common British *Helices* to a close.

Two little shells, pretty generally distributed, *Azeca tridens* and *Zua lubrica*, ought not to pass unnoticed. The first named, a link between *Bulimus* and *Clausilia*, may be found amongst herbage and on damp moss; the latter under stones and logs in moist situations. The former has the mouth furnished with teeth and folds, the outer lip notched, and the inner lip thickened; the latter possesses exactly opposite characters.

CHAPTER VI.

THERE are but few species to add to the list of mollusca inhabiting the chalk. Owing to their minute size, some of them are very difficult to procure; and perhaps the readiest way to obtain specimens is to take out a few small linen bags and an old newspaper, and in dry weather to pull up moss, grass, &c., and shake out the sand and earth from the roots on to the paper. This may then be put into a bag with a memorandum of the locality whence taken, and, being carefully tied up, may be carried home and examined at leisure. In this way have been obtained *Zonites crystallinus*, *Helix pygmœa* and *aculeata*, *Acme lineata*, and different species of *Vertigo*, which otherwise might never have been procured.

Helix pygmœa (Pl. IX., fig. 11) has been also taken successfully, and in some numbers, by sweeping the wet grass and herbage after rain with an entomologist's gauze net; and Dr. Turton found that a handful of dead and moist leaves, after being spread out on paper to dry, yielded a good harvest of these small mollusca.

One of the handsomest land-shells, and one which is partial to chalky and calcareous soils, is *Cyclostoma elegans* (Pl. VIII., fig. 8). In shape it is something between *Helix* and *Bulimus*, and in colour a yellowish-brown, with more or less of a reddish tinge, marked with purple blotches. The whorls are very finely striated in the direction in which they curve, and the sutures between the whorls very deep. The entire shell is about six-tenths of an inch in length, by four-tenths at the widest part. In Suffolk a variety has been found of a uniform buff or clay colour, without the beautiful purple spots which generally ornament the shell in specimens from the south of England (*cf.* King, " Zoologist," 1853, p. 3916).

The mouth of this shell is closed with a very solid operculum, covered on both sides with a thick epidermis, a double fringe of which completely encircles it, and causes it to appear laminated. The animal itself is of very shy and retiring habits, and in dry weather buries itself in the earth, where

it often falls a prey to carnivorous beetles, notwithstanding its closely-fitting operculum. This is not
the only mollusk which furnishes food to beetles.
We have frequently found these insects upon
partially devoured remains of different species of
Helix; and entomologists are aware that the larvæ
of some beetles are constantly found in empty snail-
shells (*cf.* "Proc. Ent. Soc." 1858, p. 9). Some years
since, when rambling over the Sussex Downs in the
spring of the year, we discovered the larva of *Drilus
flavescens* in an old shell of *Helix ericetorum.*
When exhibited some time afterwards at a meeting
of the Entomological Society, it was considered to
be but the second female example which had been
obtained in this country (*cf.* "Zoologist," 1868,
p. 1137). Coleopterists from this may take a hint.

The epiphragm, where it is found in the *Helicidæ,*
is a thin plate accurately fitting the mouth of the
shell, and secreted by the animal for its protection
during periods of inactivity, but cast off at will when
no longer required. The operculum, as in the last-
named species, is a plate of a horny character permanently attached to the back of the animal's foot,
and naturally closes the aperture of the shell when

its inhabitant has retired within it. But the next genus we have to notice, *Clausilia*, is characterized by the possession of an apparatus for partially closing the last whorl of the shell within the mouth, which is neither epiphragm nor operculum. It consists of a valve (*clausilium*) attached to the pillar of the shell by an elastic hinge, and when the animal wishes to protrude itself, according to Dr. Gray (*cf.* " Zool. Journ." i., p. 212), it pushes the plate on one side into a groove situated between the inner plate and the columella, or pillar, where it is detained by the pressure of the body of the animal, leaving the aperture free ; and when the animal withdraws itself, the plate springs forward by the elasticity of its pedicle and closes the aperture.

Of this singular genus we have already noticed (p. 66) three species as being found upon the London clay. A fourth, *Clausilia rolphii* (Pl. VIII., fig. 10), may now be named as peculiar to chalky and calca-reous soils. It is larger than *C. rugosa* (or *nigricans*), but smaller than *laminata* (*bidens* of Gray) and *bipli-cata*. It differs from the first named in being more ventricose, and in having coarser striæ and a larger and broader mouth ; it is also lighter in colour. It

has been found as near London as Charlton, but it
seems to be very local, and is apparently restricted
to the southern counties of England. Other recorded
localities for it are Ashford, Sevenoaks, Southborough,
Tunbridge Wells, Coghurst Wood, Hastings, Mickleham, near Dorking, Uppark, and Buriton, near
Petersfield, Folkestone, Birdlip and Cooper's Hill,
Gloucestershire, and Charlton King's, near Cheltenham. It is generally found in damp situations in
woods, amongst dead leaves and moss, and under
nettles and dog's mercury, as well as on the trunks
of trees.

While walking over the downs, which in many
places are studded with junipers and yews, search
may be made at the roots of these trees for some
of the chrysalis shells, *Pupa secale* (or *juniperi*) and
marginata,* as well as for one or two of the whorl
shells, *Vertigo antivertigo* (Pl. II., fig. 5), and
pygmæa (Pl. VIII., fig. 3), all of which affect dry
and barren situations, and may be found in old
chalk pits, attached to chalk stones and beneath
loose flints scattered over the downs. We have
already mentioned *Balia perversa*, but may refer to

---

* *Pupa marginata* is not uncommon at Hampton (Rich).

it again here to note that it is not uncommon beneath the bark of old hawthorns on the downs, and may be found also on old willows and ash trees, and on walls beneath ivy.

The true slugs, which remain to be noticed, do not possess the same interest for collectors as the snails; for, although they are not, strictly speaking, shell-less, their shells are comparatively insignificant, while the repulsive appearance of the animals frequently causes them to be purposely avoided.

The popular belief that slugs are only snails, which for some reason best known to themselves have temporarily vacated their shells, is not confirmed by the observations of malacologists, who, on the contrary, have long been aware of the fact that slugs are really possessed of shells of their own—of which, to say the least, they would find it extremely difficult to divest themselves.

We have already alluded to the prominence on the back of the slug; this is the mantle, on the under surface of which the shelly secretion is spread out in the form of a thin plate, and the latter from its position, is no doubt designed as a protection to the important organ beneath it. This plate, as may be

anticipated, varies in the different species; and in
the so-called shell-slug (*Testacella*), as we have
already pointed out, it becomes external, and is
situated on the back near the tail.

The common species are the black slug (*Arion
ater*), the garden slug (*A. hortensis*), the great slug
(*Limax maximus*), the yellow slug (*L. flavus*), and
the field slug (*L. agrestis*). The two first-named
are very destructive to plants and fruit, and are also
carnivorous, feeding on earth worms and decaying
animal matter, and sometimes even indulging in
cannibalism. The *Limax maximus* and *L. flavus*
frequent our houses and stables, and assiduously take
upon themselves the duties of scavengers during the
hours of darkness, retiring to out-of-the-way damp
places and drains in the day time. They do not,
however, confine themselves to buildings, but may
be found also under stones in damp situations and
about decaying stumps in woods.

The *Limax agrestis* is the pest of the farmer as
well as the gardener, and may be met with in the
corn-fields in such abundance that even the willing
rooks and pheasants find it no easy task to keep its
numbers within safe limits. It is stated in Bell's

" British Quadrupeds," 2nd. ed. p. 107, that the
small field slug, *Limax agrestis*, is a favourite morsel
with the hedgehog ; and is often scratched out and
eaten in the summer months when concealed in
the day in crevices, or amongst the roots of
grass or other close herbage. This slug also furnishes food to the blind worm, *Anguis fragilis*, which
seizes it as a dog would seize a rat, and, after holding
it for some time in its mouth, passes it slowly
through its jaws and swallows it head foremost.

The eggs of the garden slug (*A. hortensis*) are
phosphorescent for about a fortnight after they have
been deposited, and may be seen in the evening on
moist hedge banks, giving out a pale light.

We do not propose to say as much of the slugs as
our opportunities have enabled us to do in the case
of the snails; and, indeed, we may admit that we have
not paid equal attention to them. There can be no
doubt, however, that, although the handling of these
slimy creatures is not particularly agreeable, the
study of their habits and economy opens a wider field
for research, with a greater prospect of novelty, than
can be expected in the case of the testaceous
mollusca.

CONCLUSION.

Hints to Collectors—When to Collect—How to Collect—And how to
Preserve Shells.

THE best time for collecting is in the autumn, when
the mollusks are full grown, and before the beauty
of the shells becomes destroyed by the winter rains.

At the commencement of the rainy season, too,
snails are much more readily discovered, since they
leave their retreat and may be seen climbing over
trees in all directions. From this habit, indeed,
they have been found to furnish some indication of
approaching change in the weather. An American
naturalist, Mr. B. Thomas of Cincinnati, has observed that, as natural barometers, snails are more
reliable than leaves; that in consequence of their
never drinking, all the moisture they receive is by
absorption of rain, mist, or dew through the tissues
of their bodies, and this they afterwards exude at
regular intervals until they obtain a fresh supply.
Two days before rain is about to fall they climb
trees, which, according to Mr. Thomas, they never
do on other occasions; and when they are observed

to leave the herbage and get on rocks, it is a certain prognostication of wet weather. He adds, that the colour of certain kinds of snail varies according to the quantity of moisture retained.

As regards the particular time of day to collect with advantage, a search in an open country should be prosecuted after a shower of rain, or during early morn. In damp woods, where throughout the day the air is sufficiently moist to maintain the animals in full activity, no such considerations determine the best time for collecting. In such places, light is usually the desideratum, and, consequently, a search at mid-day in a clear sky is most likely to prove remunerative.

Land shells are most abundant on limestone soils, which are most congenial to their existence and perpetuation; and in explanation it may be observed that the shell is composed almost entirely of carbonate of lime; that the plants upon which the animals feed are the sources whence the mineral matters are derived, and that plants affecting calcareous soils contain proportionately a larger amount of salts than those inhabiting clayey or sandy soils. These, therefore, are in greater request. So also

many species of horse tails and grasses, which contain a large percentage of earthy salts, are on this
account much frequented by land snails.

As to the mode of preparing specimens for the
cabinet, the animals may be killed with boiling
water and removed from their shells by the aid of
a bent pin. Those which retire too far to be reached
by this ordinary expedient, as *Clausilia, Bulimus,*
and others, may be killed by placing them in tepid
water, and adding gradually hot water; the animals
may then be partially removed. The shells should
be well dried, to remove the moisture and harden
the soft parts remaining; but the heat must not be
too great, or else the shells will be discoloured, and
are liable to be broken ; and, further, the specimens
should be well dried before placing them in the cabinet, which should be in a well ventilated place, free
from damp ; for their freshness and beauty are apt
to be lost by the growth of fungi upon their surfaces.

In cleaning the shells of some species, great care
is needed, so as not to remove the hairs or bristles
which clothe the surface of the epidermis.

The shells of such snails as *Paludina, Cyclostoma,*
and others, would be imperfectly illustrated without

the opercula which close the apertures of their shells. Each one should be detached from the foot of the snail, the interior of the shell plugged with cotton wool, and the specimen gummed down in its natural position.

The shells of mussels and other bivalves which gape a great deal after the animal has been removed, should be carefully closed and bound with thread until dry. Bivalves as small as *Sphærium corneum* may be treated in this way, but the smaller species of *Pisidium* and some of the smaller univalves, as, for instance, the little Sedge-shell *Carychium minimum*, may be dried in hot sand. Care, however, is required in the process, since too much heat will cause a transfusion of the carbonaceous matter of the animal into the substance of the shell, and so discolour it.

Slugs require a different treatment. On this subject, Mr. Tate, in his " British Land and Freshwater Mollusca," says : " As regards the internal shell, it may be obtained by making a conical incision in the shield, taking care not to cut down upon the calcareous plate, which can then be removed without difficulty. The animals can only be conserved by keeping them in some preservative fluid ; but the great object to keep in view is to have the slug naturally extended. Most fluids contract the slugs when they are immersed in them. The slugs should be killed whilst crawling, by plunging them into a solution of corrosive sublimate, or into benzine. Models in wax or dough are sometimes substituted for the animals. A writer in the "Naturalist"* gives a process for the preservation of slugs, which he states to answer admirably, and to be very superior to spirit, glycerine, creosote, and other solutions :— "Make a cold saturated solution of corrosive sublimate ; put it into a deep wide-mouthed bottle, then take a slug you wish to preserve and let it crawl on a long slip of card. When the tentacles are fully extended, plunge it suddenly into the solution ; in a few minutes it will die, with the tentacles fully extended in the most lifelike manner, so much so, indeed, that if taken out of the fluid it would be difficult to say whether it be alive or dead. The slugs thus prepared should not be mounted in spirit, as it is apt to contract and discolour them. A mixture of one and a half parts of water and one part of glycerine has been found to be the best mounting fluid ; it preserves the colour

* The "Naturalist," vol. i., p. 253 (1865).

beautifully, and its antiseptic qualities are unexceptional. A good-sized test tube answers better than a bottle for putting them up, as it admits of closer examination of the animal. The only drawback to this process is, that unless the solution is of sufficient strength, and unless the tentacles are extruded when the animal is immersed, it generally, but not invariably, fails. Some slugs appear to be more susceptible to the action of the fluid than others; and it generally answers better with full-grown than with young specimens. But if successful, the specimens are as satisfactory as could be desired; and even if unsuccessful, they are a great deal better than those preserved in spirit; for although the tentacles may not be completely extruded, they are more or less so."

The *Testacellæ*, or shell-slugs, may be preserved by partially drying them in sand and removing the soft parts through a slit in the length of the foot, filling up with cotton wool, and completing the drying process.

As to the best mode of exhibiting shells in the cabinet, opinions differ. They may be gummed on slips of card, kept loose in card-board trays, or, better still, in glass-topped card-board boxes. We much prefer the last-named method for many reasons. Being thus protected from dust, they retain their natural appearance better ; nearly allied species, or specimens of the same species from different localities, may be distinguished without risk of their getting mixed, and the bottom of each box serves as a tablet whereon to pencil the collector's notes.

The boxes may be arranged systematically, or in any way most convenient to the collector. For the benefit of those who may adopt the former and more desirable method, we subjoin a systematic list of all the British Land and Freshwater Mollusca, placing opposite to each of the species referred to in this book, the number of the page whereon it is mentioned.

In conclusion, we shall add a catalogue of such works or published articles as we have met with which contain notices of the land and freshwater shells of particular districts. Such publications are always of interest to local collectors ; and, as indicating the source of much information on the subject of our native mollusca, we believe that a list of them will be useful to conchologists.

SYSTEMATIC LIST

OF THE

BRITISH LAND AND FRESHWATER MOLLUSCA.

I.—AQUATIC.

BIVALVES.

(Conchifera.)

Lamellibranchiata.

FAM. SPHÆRIIDÆ.

Sphærium corneum, 42.
„ rivicola, 42.
„ ovale, 43.
„ lacustre, 43.
Pisidium amnicum, 43.
„ fontinale, 43.
„ pusillum, 43.
„ nitidum, 43.
„ roseum, 43.

FAM. UNIONIDÆ.

Unio tumidus, 35, 36.
„ pictorum, 35, 36.
„ margaritifer, 35.

UNIVALVES.

(Gasteropoda).

Pectinibranchiata.

FAM. NERITIDÆ.

Neritina fluviatilis, 18, 55.

FAM. PALUDINIDÆ.

Paludina contecta, 13, 55.
„ vivipara, 13, 55.
Bythinia tentaculata, 52.
„ leachii, 52, 53.
Hydrobia similis, 52, 53.
„ ventrosa, 54.

FAM. VALVATIDÆ.

Valvata piscinalis, 51, 56.
„ cristata

II.—TERRESTRIAL.

UNIVALVES.

(Gasteropoda.)

FAM. LIMACIDÆ.

Arion ater, 90.
  „   *hortensis,* 90, 91.
Geomalacus maculosus
Limax gagates
  „   *marginatus*
  „   *flavus,* 90.
  „   *agrestis,* 90, 91.
  „   *arborum*
  „   *maximus,* 90.

FAM. TESTACELLIDÆ.

Testacella haliotidea, 31.

FAM. HELICIDÆ.

Succinea putris, 57, 61.
  „   *elegans,* 57, 61.
  „   *oblonga,* 61.
Vitrina pellucida, 62.
Zonites cellarius, 6, 28.
  „   *alliarius,* 29.
  „   *nitidulus,* 6, 29.
  „   *purus*
  „   *radiatulus,* 5.
  „   *nitidus,* 6, 29.
  „   *excavatus*
  „   *crystallinus,* 29.
  „   *fulvus*

Helix lamellata
  „   *aculeata,* 82, 83.
  „   *pomatia,* 68, 73.
  „   *aspersa,* 19, 23.
  „   *nemoralis,* 23, 25.
  „   *arbustorum,* 23.
  „   *cantiana,* 79, 80.
  „   *cartusiana,* 57, 80.
  „   *rufescens,* 25, 26.
  „   *concinna,* 27.
  „   *hispida,* 5, 26.
  „   *sericea*
  „   *revelata*
  „   *fusca*
  „   *pisana,* 76, 77.
  „   *virgata,* 75, 76.
  „   *caperata,* 75, 77.
  „   *ericetorum,* 76, 86.
  „   *rotundata,* 5, 27, 83.
  „   *rupestris*
  „   *pygmœa,* 84.
  „   *pulchella,* 28.
  „   *lapicida,* 78.
  „   *obvoluta,* 81.
Bulimus acutus, 14.
  „   *montanus,* 14, 64, 65.
  „   *obscurus,* 14, 64.

Fam. Helicidæ—*contd.*

Pupa secale, 14, 88.

„   *ringens*, 14.

„   *umbilicata*, 14, 30.

„   *marginata*, 14, 30, 88.

Vertigo antivertigo, 88.

„   *pygmæa*, 88.

„   *alpestris*

„   *substriata*

„   *pusilla*

„   *angustior*

„   *edentula*

„   *minutissima*

Balia perversa, 64, 88.

Clausilia rugosa, 6, 15, 67.

Fam. Helicidæ—*contd.*

„   *rolphii*, 15, 67, 87.

„   *biplicata*, 15, 66.

„   *laminata*, 15, 67, 87.

Azeca tridens, 83.

Zua lubrica, 83.

Achatina acicula, 62.

Fam. Carychiidæ.

Carychium minimum, 11, 63.

Fam. Cyclostomatidæ.

Cyclostoma elegans, 11, 18, 85.

Acme lineata, 84.

LIST OF LOCAL CATALOGUES

OF

LAND AND FRESHWATER MOLLUSCA.

ABERDEEN, KINCARDINE, AND BANFF.—*Macgillivray*, History of Mollusca of these Counties, 8vo, 1843; and *Taylor*, Zoologist, vol. xi., p. 3878.

ARRAN, ISLE OF.—*Landsborough*, Mollusca of Whitney Bay; Zoologist, vol. i., p. 86.

BERWICKSHIRE AND NORTH DURHAM.—*Johnston*, Trans. Berw. Nat. Hist. Soc., 1838, p. 154; Trans. Tyneside Nat. Club, vol. i., p. 97 (1850).

CHESHIRE.—*Bellairs*, Land and Freshwater Shells in the vicinity of Chester. Pamphlet, n.d.

CORNWALL.—*Borlase*, Nat. Hist. Cornwall, folio, 1758; *Cocks*, Contrib. Fauna Falmouth, 1845; and *King*, Freshwater Shells of Cornwall, Zoologist, vol. xii., pp. 1038 and 1194.

DERBYSHIRE.—*Bloxham*, Mag. Nat. Hist., vol. vi., p. 324.

DEVONSHIRE.—*Bellamy*, Cat. Land and Freshwater Shells found in the vicinity of Plymouth, Edinb. Journ. Nat. Hist., Oct. 1837, p. 115; *Parfitt*, List Land and Freshwater Shells in neighbourhood of Exeter, Naturalist, vol. iv., p. 150 (1854).

DORSETSHIRE.—*Pulteney*, Catalogue Birds, Shells, &c., of Dorset, folio, 1799, and the "Ornithology and Conchology of the County of Dorset," by J. C. Mansel Pleydell. Privately printed, 1874, *vide* "Athenæum," 25th July, 1874, p. 118.

DURHAM.—Tyneside Nat. Club Trans., vol. i., p. 97 (1850).

ESSEX.—*Dale*, Nat. Hist. Harwich, 4to, 1732.

GLOUCESTERSHIRE.—*Webster*, Land and Freshwater Mollusks found near Cheltenham, Naturalist, vol. iv., p. 175 (1854); *Prentice*, Ann. Mag. Nat. Hist., 1856, p. 446.

IRELAND.—*Thompson*, Catalogue of Mollusca, Ann. Mag. Nat. Hist., vol. vi., pp. 16, 109, 194 (1841); *Walpole*, Cat. Shells, Dublin, Zoologist, vol. xi., p. 4022; and *Clarke*, on Genus *Limax* in Ireland. Ann. Mag. Nat. Hist., 1843, p. 332.

ISLE OF MAN.—*Forbes*, Mag. Nat. Hist., vol. viii., p. 69; and Malacologia Monensis, 8vo, 1838, p. 63.

ISLE OF WIGHT.—*Guyon*, Cat. Land and Freshwater Shells of the Island, in Venables' Guide to the Isle of Wight, pp. 461–465.

KENT.—*Smith*, Mollusca of Levendale, Zoologist, vol. xii., p. 4332; and *Benson*, Ann. Mag. Nat. Hist., 1856, p. 74.; *R. H. S. Smith*, List Land and Freshwater Mollusks found near Sevenoaks, and Mollusca in vicinity of Faversham, Cassell's Floral Guide to East Kent, p. 63 (1839).

LANCASHIRE.—*Leigh*, Nat. Hist. Cheshire, Lancashire, and the Peak, folio, 1700; and *Kenyon*, Shells in Neighbourhood of Preston, Mag. Nat. Hist., vol. ii., pp. 273 and 303; and Report of Bury Nat. Hist. Soc. (1872).

LINCOLNSHIRE.—*Ball*, Land Shells of Lincolnshire, "Young England," May 1, 1864, p. 76.

LONDON AND ENVIRONS.—*Gray*, Ann. Mag. Nat. Hist., 1856, pp. 465 and 25; *Cooper*, Appendix to Flora Metropolitana, 12mo, 1836, pp. 120–127; and *Sheppard*, Mollusca of Fulham, Zoologist, vol. ix., p. 3120.

MIDDLESEX.—*See* London and Environs.

MORAY.—*Gordon*, List of Mollusca of Moray and Moray Firth, Zoologist, vol. xii., pp. 4318, 4421, 4443.

NORFOLK.—*Bloxam*, Mag. Nat. Hist., vol. vi., p. 324; *Bridgeman*, Mollusca of Norwich, Zoologist, vol. viii., p. 2741, and vol. ix., p. 3302; and Trans. Norf. and Norw. Nat. Soc., 1871–72, p. 47.

NORTHAMPTONSHIRE. — *Morton*, Nat. Hist. Northampton, folio, 1712; and Mollusca of Mears Ashby, Phil. Trans., vol. xxv., p. 325.

NORTHUMBERLAND. — *Alder*, Trans. Nat. Hist. Society, Northumb., 1830, p. 16 ; suppl., 1833, p. 5; Tyneside Nat. Club Trans., vol. i., p. 97 (1850).

NOTTINGHAMSHIRE.—*Lowe*, Zoologist, vol. x., p. 3390.

OXFORDSHIRE.—*Plot*, Nat. Hist., Oxford, folio, 1676; *Norman*, Zoologist, vol. xi., p. 4126; *Streech*, Mollusca of Banbury, Zoologist, vol. xiii., pp. 4540 and 4658; *J. D.*, Shells in vicinity of Oxford, Naturalist, vol. v., p. 200 (1855); *Whiteaves*, Mollusca near Oxford, 8vo, 1857, pp. 20; and *Stubbs*, Mollusca in vicinity of Henley-upon-Thames, Zoologist, vol. iv., 2nd series, p. 1836.

SHROPSHIRE.—*Gwyn Jeffreys*, Ann. Mag. Nat. Hist., 1855, p. 464.

SOMERSETSHIRE.—*Miller*, Shells of Bristol and Environs, Ann. Phil., 2nd series, vol. vii., p. 376 (1822); *Norman*, Proc. Somers. Archæol. and Nat. Hist. Soc., 1860; pub. separately, 1861, 8vo, pp. 23; *Jellie*, Mollusca of Bristol, Naturalist, vol. iii., p. 148 (1867).

STAFFORDSHIRE.—*Plot*, Nat. Hist. Stafford, folio, 1686; and *Sir O. Mosley*, Nat. Hist. Tutbury, 1863.

SUFFOLK.—*Sheppard*, Trans. Linn. Soc., 1825, p. 148; and *King*, Shells near Sudbury, Zoologist, vol. xi., p. 3913.

SURREY.—*Cooper*, Shells at Mickleham, Mag. Zool. and Bot., vol. ii., p. 471; and *Saunders*, List of the Mollusca of Reigate, 2nd ed., 1864, p. 23.

SUSSEX.—*Unwin*, List Land and Freshwater Mollusca in vicinity of Lewes, Naturalist, vol. iii., p. 54 (1853); *Merrifield*, Sketch Nat. Hist. Brighton, 8vo, 1864, pp. 155–161; *Chambers*, Handbook for Eastbourne, 1873.

WARWICKSHIRE.—*Nelson*, *Limnæidæ* of Birmingham, Naturalist, vol. iii., p. 26 (1867); and *Tye*, Proc. Birm. Nat. Hist. Soc., Part I., pp. 106–109 (1869).

WEXFORD.—*Hanley*, Freshwater Shells of Wexford, Ann. Mag. Nat. Hist., vol. vi., p. 395.

WORCESTERSHIRE.—*Griffiths*, Cat. Land and Freshwater Mollusca of the Malvern District, Trans. Malvern Nat. Field Club, Part. III. (1870), pp. 157–166.

YORKSHIRE.—*Hincks*, Shells of York and Yorkshire, Ann. Mag. Nat. Hist., vol. iii., p. 366; *Nunneley*, on Species of the Genus *Limax* near Leeds, Trans. Phil. Soc. Leeds, 1837, p. 41; *Ashford*, Mollusca of Ackworth, Zoologist, 1854, p. 4261; *Watson*, Naturalist, vol. iv., p. 85 (1854); *Bean*, Mollusca of Scarborough, in Theakston's Guide to Scarborough, pp. 153–158 (1862); *Blackburn*, Mollusks collected at Knaresborough, Naturalist, vol. iii., p. 98 (1867); *Stevenson*, Yorkshire Naturalists' Recorder, 1872; and *Hebden*, List of Shells in neighbourhood of Wakefield, Quart. Journ. Conchol., 1874, p. 3.

INDEX.

Woodfall & Kinder, Printers, Milford Lane, Strand, London, W.C.